Engendering Climate Change

This book focuses on the gendered experiences of environmental change across different geographies and social contexts in South Asia and on diverse strategies of adapting to climate variability.

The book analyzes how changes in rainfall patterns, floods, droughts, heatwaves and landslides affect those who are directly dependent on the agrarian economy. It examines the socio-economic pressures, including the increase in women's work burdens both in production and reproduction on gender relations. It also examines coping mechanisms such as male migration and the formation of women's collectives which create space for agency and change in rigid social relations. The volume looks at perspectives from India, Pakistan, Bangladesh and Nepal to present the nuances of gender relations across borders along with similarities and differences across geographical, socio-cultural and policy contexts.

This book will be of interest to researchers and students of sociology, development, gender, economics, environmental studies and South Asian studies. It will also be useful for policymakers, NGOs and think tanks working in the areas of gender, climate change and development.

Asha Hans is a former Director of the School of Women's Studies and Professor of Political Science, Utkal University, India.

Nitya Rao is Professor of Gender and Development at the School of International Development, University of East Anglia, UK.

Anjal Prakash is the Research Director and Adjunct Associate Professor, Bharti Institute of Public Policy, Indian School of Business, Hyderabad, India.

Amrita Patel has been an adviser in the Department of Women and Child Development, Government of Odisha, India.

Engendering Climate Change
Learnings from South Asia

Edited By Asha Hans, Nitya Rao, Anjal Prakash and Amrita Patel

Routledge
Taylor & Francis Group
LONDON AND NEW YORK

First published 2021
by Routledge
2 Park Square, Milton Park, Abingdon, Oxon OX14 4RN

and by Routledge
52 Vanderbilt Avenue, New York, NY 10017

*Routledge is an imprint of the Taylor & Francis Group, an
informa business*

British Library Cataloguing-in-Publication Data
A catalogue record for this book is available from the British
Library

Library of Congress Cataloging-in-Publication Data
A catalog record has been requested for this book

ISBN: 978-0-815-36165-7 (hbk)
ISBN: 978-0-367-69586-6 (pbk)
ISBN: 978-1-003-14240-9 (ebk)

Typeset in Sabon
by SPi Global, India

Contents

List of figures   vii
List of tables   ix
List of contributors   x
Foreword   xvii
Acknowledgements   xix

1 Gender, climate change and the politics of vulnerability: an
   introduction   1
   NITYA RAO, ANJAL PRAKASH, ASHA HANS, AND AMRITA PATEL

PART I
Vulnerabilities   17

2 Vulnerabilities of rural women to climate extremes: a case of
   semi-arid districts in Pakistan   19
   AYESHA QAISRANI AND SAMAVIA BATOOL

3 Gendered vulnerabilities in *Diaras*: struggles with floods in the
   Gandak river basin in Bihar, India   38
   PRANITA BHUSHAN UDAS, ANJAL PRAKASH, AND CHANDA GURUNG GOODRICH

4 Of borewells and bicycles: the gendered nature of water access in
   Karnataka, South India and its implications for local vulnerability 58
   CHANDNI SINGH

5 Vulnerabilities and resilience of local women towards
   climate change in the Indus basin   85
   SAQIB SHAKEEL ABBASI, MUHAMMAD ZUBAIR ANWAR, NUSRAT HABIB,
   AND QAISER KHAN

6 Climate change, gendered vulnerabilities and resilience in high
   mountain communities: the case of Upper Rasuwa in Gandaki
   River Basin, Hindu Kush Himalayas   106
   DEEPAK DORJE TAMANG AND PRANITA BHUSHAN UDAS

PART II
Adaptation and Wellbeing   125

7 Wells and well-being in South India: gender dimensions of
   groundwater dependence   127
   DIVYA SUSAN SOLOMON AND NITYA RAO

8 Gender, migration and environmental change in the Ganges-
   Brahmaputra-Meghna delta in Bangladesh   152
   KATHARINE VINCENT, RICARDO SAFRA DE CAMPOS, ATTILA N. LÁZÁR, AND
   ANWARA BEGUM

9 Women-headed households, migration and adaptation to climate
   change in the Mahanadi Delta, India   172
   SUGATA HAZRA, AMRITA PATEL, SHOUVIK DAS, ASHA HANS, AMIT GHOSH, AND
   JASMINE GIRI

10 Gender dynamics and climate variability: mapping the linkages
    in the Upper Ganga Basin in Uttarakhand, India   201
    VANI RIJHWANI, DIVYA SHARMA, NEHA KHANDEKAR, ROSHAN RATHOD, AND
    MINI GOVINDAN

11 Shaping gendered responses to climate change in South Asia   226
    ASHA HANS, ANJAL PRAKASH, NITYA RAO, AND AMRITA PATEL

Index   236

Figures

2.1	Temperature trend for Faisalabad	24
2.2	Temperature trend for Dera Ghazi Khan	25
3.1	Flood zones in Bihar	40
3.2	Pipra-Piprasi Embankment in Gandak River	42
3.3	Disparities in Landownership in West Champaran	42
3.4	Top down versus bottom up approach of vulnerability	44
3.5	Differential gender impact of climate change in an agrarian society	45
3.6	Loan burden among households in a village in Nautan	49
4.1	Map of study area: Kolar District in South India showing research sites	63
4.2	Block-wise rainfall patterns in Kolar District, 1900–2005 (top); average annual rainfall in Kolar, 2006–2014 (bottom)	64
4.3	Main employment categories in Kolar	65
4.4	Types of land use in Kolar (left) and area under different types of cultivation (right)	67
4.5	Key risks to agriculture ranked by men (top) and women (bottom)	71
4.6	Principal source of drinking water in Kolar	74
4.7	Plastic pots for water – public taps exist but are not reliable and involve long waiting times	75
5.1	HI-AWARE intervention sites in Pakistan	88
5.2	Women's perception: Climate induced events in Upstream Basin	92
5.3	Women's perception: Climate-induced events in the High Rainfall Zone MIB (Pakistan)	95
5.4	Women's perception: Climate-induced events in the Medium Rainfall Zone MIB (Pakistan)	96
5.5	Women's perception: Climate-induced events in the Low Rainfall Zone MIB (Pakistan)	96
5.6	Women's perception: Climate-induced events in the Low Indus Basin LIB (Pakistan)	97

6.1   Gandaki river basin   108
6.2   Satellite image of study villages in Rasuwa district   109
7.1   Location of Coimbatore District in India   129
7.2   Plantation area of dryland crops (coarse millets and grains)
       in Coimbatore District, 1997–2015   133
7.3   Average pre-monsoon (May) and post-monsoon (July)
       groundwater levels in Coimbatore District, 1997–2015   140
8.1   Gendered migration patterns from the Ganges-Brahmaputra-
       Meghna delta in Bangladesh   158
8.2   Reason for migration in migrant households   160
8.3   Percentage of migrant households receiving remittances   161
8.4   Perceived success of migration by women   162
8.5   Level of happiness of women in migrant and non-migrant
       households   163
8.6   Percentage of women in migrant households who intend to
       migrate in the future   163
8.7   Barriers to migration for women in non-migrant and migrant
       households   164
9.1   The study area map of Mahanadi Delta   175
9.2   Multi-hazard risk and migration map of Mahanadi Delta   179
9.3   Major components in vulnerability assessment   186
9.4   Distribution of households in five multi-hazard zones   187
9.5a Dimensions of migration at individual level (migrant)   190
9.5b Dimensions of migration at collective level (household
       and village)   190
10.1 Conceptual framework of the study   203
10.2 Yearly average temperature trend in Kim Khola, Hakeempur
       Turra, and Huddu respectively for 1901–2015   208
10.3 Water Resource Map for Hakeempur Turra Village   210
10.4 Map depicting the study sites in the high, mid- and plain
       regions of Uttarakhand   215

Tables

2.1	Sample selection by district	21
2.2	Socio-economic indicators by district	23
4.1	Changing cropping patterns and practices in Kolar	66
4.A.1	Rationale for choice of blocks in Kolar district	84
4.A.2	Details of villages studied in Kolar	84
5.1	Socio-economic situation of women in Indus	90
5.2	Women's vulnerabilities to climate change in Indus basin	99
6.1	Broad task allocation study (TAS) in the case study sites	114
7.1	Land and well ownership among households sampled by caste	131
7.2	Intra-household strategies for coping with water shortages in order of preference	144
9.1	Socio-economic profile of the five districts of the Mahanadi Delta	177
9.2	Socio-economic characteristics of respondents	181
9.3	Vulnerability Assessment of Men and Women Headed Households in MD	184
9.4	Patterns of Migration from Women Headed Households in MD	188
9.5	Adaptation Activities of Women Headed Households in the last 5 years	192
10.1	Studied biophysical and community assets	204
10.2	Study sites	206
10.3	Perceived climatic variability in plains elevation site	208
10.4	Perceived climatic variability in mid-elevation site	212
10.5	Perceived climatic variability in high-elevation site	213
10.6	Distribution of potable water source in Hakeempur Turra	214
10.7	Rules and implications for actors in action situation 1: accessing water for domestic use	216
10.8	Rules and implications for actors in the high hills, mid-hills and plains in action situation 2: agriculture-related activities	218

Contributors

Saqib Shakeel Abbasi is Scientific Officer at the Social Sciences Research Institute, Pakistan Agricultural Research Council, Islamabad, Pakistan. He holds a master's degree in social sciences from SZABIST, Islamabad and a master's degree in food security and agricultural development from Kyungpook National University, South Korea. He has approximately ten years' experience of working in public-sector organizations in the fields of agricultural research and social development. He has worked on Pakistan–China cooperation projects for Pakistan's APEX research organization, leading operations and administration in the agribusiness and marketing department. He has also been actively involved in different climate change and food security-related research and development projects. His main interest is in developing appropriate solutions for ensuring sustainable livelihoods in marginal areas of the country, with a special focus on mountain communities.

Muhammad Zubair Anwar is Principal Scientific Officer, Social Sciences Research Institute at Pakistan Agricultural Research Council, Islamabad, Pakistan. He has a master's degree in rural sociology from the University of Agriculture, Faisalabad and a PhD in anthropology from Quaid-i-Azam University, Islamabad. During 25 year of service, Dr Zubair has worked on various national and international projects. In the course of his career he has produced about 63 research reports and 48 research publications/articles in national and international journals. As a programme leader, he has supervised the research activities of his fellow scientists and students. He has also had experience of working with national and international organizations including ICARDA, ACIAR-CIMMYT, IRRI and ICIMOD.

Samavia Batool holds a master's degree in economics from Quaid-i-Azam University, Islamabad. She works as a Senior Research Associate with the Sustainable Development Policy Institute, Islamabad.

Anwara Begum is a Senior Research Fellow at the Bangladesh Institute of Development Studies, and at present she is also the Division Chief of the Human Resource Development Division. She is a Board Member of several NGOs. In her book, "Destination Dhaka" she has delved into the heterogeneity of poor pavement and slum dwellers' migration commitments, conceptualizing a new theoretical framework of migration. She has extensive experience researching migration in developing countries, including Bangladesh. She worked on the first MDG Global report, PRSP I and II; Bangladesh Government reports, UNDP Global Follow-up, Unifem Female Migration and EFA reports etc. She has worked for three decades on Urbanization, Migration, Gender, Housing, Education, Poverty, Health, Water and Sanitation, Evaluation and Monitoring. She was Director Research at CIRDAP in the Dhaka headquarters, which represents the 15 CIRDAP Member Countries and their related Ministries. Recently, she worked as Gender Specialist and Focal Point, Resettlement Expert, for DECCMA, Lead Researcher for Labour Market and Skills Gap Report on Nurses for ADB, and published reports on Adapting to Climate Change and Inclusive Care of the Elderly in Bangladesh.

Shouvik Das is a Research Scholar at the School of Oceanographic Studies, Jadavpur University, India and has completed an MSc in geography at Calcutta University and an MPhil in oceanography and coastal management at Jadavpur University. He has worked as a Junior Research Fellow (SIS-DP project) in DST, Government of West Bengal. India, and as a Doctoral Fellow (DECCMA project) at Jadavpur University. His research interests are climate change, vulnerability, migration, adaptation, geo-informatics, social surveys and statistical data analysis.

Amit Ghosh is a Research Scholar at Jadavpur University School of Oceanographic Studies working on vulnerability, hazards and climate-change hotspot mapping. He gained a master's degree in remote sensing and geographic information systems after a bachelor's degree in geography. His basic interests are in geospatial science, distributed hydrology and agroecology, natural hazards and climate change.

Jasmine Giri has an MA and MPhil in sociology from Sambalpur University, Odisha. She has researched on migration and women, MGN-REGS and child nutrition, and has managed projects on livelihoods. She is a Research Fellow on the DECCMA Project on Deltas and Climate Change.

Chanda Gurung Goodrich is the Senior Gender Specialist – Gender Lead at the International Center for Integrated Mountain Development (ICIMOD), in Kathmandu, Nepal. She holds an MPhil and PhD from the School of International Relations in Jawaharlal Nehru University, New Delhi, India. Her previous work experience is as Principal Scientist with the International Crop Research Institute for the Semi-Arid Tropics (ICRISAT), India and as Executive Director with the South Asia Consortium for Interdisciplinary Water Resources Studies (SaciWATERs), India. Dr Goodrich's professional interest is in gender research in natural resource management, agriculture and livelihoods, especially for small-holder farmers.

Mini Govindan, Fellow at the Energy and Resources Institute (TERI), New Delhi, has a PhD in development studies from the Institute for Social and Economic Change, Bangalore, India. She has led and been part of several research projects on the themes of energy and gender, social inclusion, climate change, natural resources management and sustainable development.

Nusrat Habib was previously a Senior Scientific Officer at the Social Sciences Research Institute, Pakistan Agricultural Research Council, Islamabad, Pakistan (PARC), where her main areas of research covered climate change, migration, rural livelihoods and gender issues. Currently she is pursuing her PhD studies at the University of Queensland, Australia. She holds an MPhil in Applied Economics from University of Agriculture, Peshawar-Pakistan, and an MSc Economics from University of the Punjab Lahore-Pakistan.

Asha Hans is Professor Emeritus at the Development Research Institute, former Professor of Political Science and the Founding Director of the School of Women's Studies at Utkal University, India. She is a member of the board of the UN Women's Peace Humanitarian Fund. Her research interests include gendered migration due to climate change, development projects and conflict. She has published extensively on gendered disability and humanitarian issues. Her recent publications include *The Gender Imperative,* co-edited with Prof Betty Reardon (Routledge 2019) and *Disability, Gender and the Trajectories of Power* (Sage 2015) and *Migration, Workers And Fundamental Freedoms: Pandemic Vulnerabilities and States of Exception in India* (Routledge 2021).

Sugata Hazra, a geologist by profession, is a Professor of coastal zone management and former Director of the School of Oceanographic Studies at Jadavpur University, India. His research interests include deltas, climate change, and biophysical and socio-economic impact. He has led several international and national research projects on the Sundarbans and coastal oceans and has more than 100 international publications, edited volumes and monographs to his credit.

Qaiser Khan is a Research Officer at Social Sciences Research Institute, Pakistan Agricultural Research Council, Islamabad, Pakistan. As a member of a team working on socio-economic and livelihood improvement strategies for rural households, he participates in the preparation of data collection tools and analysis; researches priority issues of national and regional importance in the agricultural social sciences; and provides research feedback to biological scientists working on crops and livestock. He has developed and strengthened linkages with research, extension, development and market agencies for the upscaling and promotion of new technologies, and has extended consultancy/advisory services to social science students.

Neha Khandekar is currently pursuing her PhD research at Ashoka Trust for Research in Ecology and the Environment (ATREE), Bengaluru. She has six years' research experience in water resources management, specifically in the area of water conflicts (around hydropower development) in Indian river basins. For the past few years, she has worked as a technical analyst in the field of water and agriculture, engaging with grassroots as well as international organizations. She holds a master's degree in water management from TERI University, New Delhi and has taken on research assignments at sub-national, national and South Asian levels. For her research, she has travelled extensively across varied mountain and plain socio-ecological systems within the catchments in the Northern Indian Himalayan region.

Attila N. Lázár is the Principal Research Fellow within the WorldPop group (Geography and Environmental Science Department) at the University of Southampton, UK. He is an environmental engineer and modeller with a hydrological background. His recent research has focused on the dynamic modelling of interlinked bio-physical and human systems, with a strong focus on deltaic environments.

Amrita Patel was a researcher, trainer and teacher of women's studies at Uktal University, India. She conducted research on climate change, land rights, displacement, migration, women's self-help groups, the declining child sex ratio and unorganized workers, among other subjects. She was a lead trainer and researcher on gender budgeting. Dr Patel died in January 2020.

Anjal Prakash is Research Director and Adjunct Associate Professor, Bharti Institute of Public Policy, Indian School of Business (ISB), Hyderabad, India. Before joining ISB, Dr Prakash briefly taught at the TERI School of Advanced Studies, New Delhi, India as an Associate Professor in the Department of Regional Water Studies. His earlier association was with the International Center for Integrated Mountain Development (ICIMOD) in Kathmandu, Nepal where he was Coordinator of the Himalayan Adaptation, Water and Resilience (HI-AWARE) research programme on glacier- and snowpack-dependent river basins. Dr Prakash was the coordinating lead author of IPCC's *Special Report on Oceans and Cryosphere in a Changing Climate* (SROCC). He is also a lead author of the chapter on cities, settlements and key infrastructure and the cross-chapter paper on mountains in the Working Group II of the IPCC's 6th Assessment Report. Dr Prakash's work focuses on gender, water and climate change, urban resilience and social inclusion issues covering South Asia.

Ayesha Qaisrani was a Senior Research Associate at the Sustainable Development Policy Institute, Islamabad. Currently, she is leading the COVID-19 and Human Rights Task Force at the Ministry of Human Rights, Islamabad as a UNDP Consultant. She is also serving as a Research Fellow for the MIGNEX project at the Lahore University of Management Sciences. She holds an MSc in migration studies from the University of Oxford and an MS in economics from the National University of Science and Technology, Islamabad. Her research interests include rural livelihoods, migration and gender issues.

Nitya Rao is Professor of Gender and Development at the University of East Anglia, UK. She has worked extensively as a researcher, practitioner and advocate in the field of women's rights, empowerment and education. Her research interests include exploring the gendered changes in land and agrarian relations, migration and livelihoods, especially in contexts of climatic variability and economic precarity, with implications for gender justice and food security. She has published extensively on these themes in international peer-reviewed journals and books. She is also a member of the High-Level Panel of Experts to the Committee on World Food Security.

Roshan Rathod, a sustainable development practitioner, has been working on issues of climate change, gender and natural resource management for over five years. She has worked on climate change adaptation from socio-cultural perspectives in varied landscapes, especially the Indian Himalayan region. She is also a trainer on gender sensitization and gender justice for various sustainable livelihood and water security stakeholders across the country. She currently works as a Research Consultant with the International Water Management Institute (IWMI), documenting best practices for springshed management across the Indian Himalayan Region. Prior to this she worked as a Gender Specialist at the People's Science Institute in Dehradun with a focus on gender, institution building and water security in the Himalayan landscape. She is a Fellow of the Ripple Academy for Women Environmental Leaders, a member of the National Facilitation Team, MAKAAM (platform for women farmers across India) and a member of Vikalp Sangam, Confluence of Alternatives, India. She has published papers, articles and op-eds in numerous online and print platforms in English, Hindi and Marathi.

Vani Rijhwani is a Research Associate at the Energy Resources Institute (TERI), New Delhi, India. She has an MPhil in natural resources management from the Indian Institute of Forest Management, Bhopal and works at the interface of climate change and sustainable development. At TERI, her work covers the arenas of international climate negotiations, MRV mechanisms, gender mainstreaming and the domestic policy narratives surrounding climate change.

Ricardo Safra de Campos is a Lecturer in human geography at the Global Systems Institute, University of Exeter, UK, focusing on mobility and migration responses to global environmental change. His research interests include spatial mobility associated with environmental factors, temporary and permanent internal mobility in developing countries, data collection methods in migration research, and sustainable livelihoods and livelihood diversification.

Divya Sharma has finished her doctoral studies at TERI School of Advanced Studies and was the recipient of a HI-AWARE Doctoral Fellowship. Her studies attempt to unravel the governance and institutional mesh by looking into social networks and decision-making processes to understand factors that lead to differential vulnerability to climate change and facilitate adaptation interventions. Her research interests include a complexity approach to the human–environment system and social networks. She has also worked as a Young Researcher from Uttarakhand for a project funded under the National Mission on Himalayan Studies. She was a DAAD scholarship recipient and a graduate exchange student at the German-Indian Sustainability and Climate Change Dialogue, Environmental Policy Research Centre (FFU) at Freie Universität Berlin. Before that, she worked for six years in the engineering and aerospace industry as an analyst.

Chandni Singh is a researcher and faculty member at the Indian Institute for Human Settlements (IIHS), Bangalore. Her research examines the human dimensions of global environmental change, focusing on drivers of differential vulnerability to climate change and hazards, linkages between climate change adaptation and development, and behavioural aspects of climate adaptation. She is a lead author on the IPCC Assessment Report 6 Working Group II, contributing author on the IPCC's Special Report on 1.5 Degrees, and serves on the editorial boards of *Regional Environmental Change*, *Climate and Development* and *Progress in Development Studies*.

Divya Susan Solomon is a PhD student in Resource, Policy, and Behavior at the University of Michigan Ann Arbor. She uses interdisciplinary methods to study changes in land and agrarian relations, migration, livelihoods, gender, food and nutrition security, and the well-being of rural communities in the global south. Previously, Divya worked with the Ashoka Trust for Research in Ecology and the Environment (ATREE) on understanding the vulnerability and adaptation of agrarian and forest-dependent communities in semi-arid regions in India.

Deepak Dorje Tamang is currently the Director of an action research NGO (SEARCH, Nepal). He has four decades of practical experience in sustainable rural development. A Senior Fellow with the World Forestry Centre, Portland, Oregon USA and an NGDO trainer facilitator with the Asian Institute of Technology (AIT), Bangkok, Thailand, he has worked with several UN agencies and bilateral international organizations including as researcher with ICIMOD. A Post-graduate Fellow in Management at Cranfield University, Bedford, UK, he was a mentor for the Queen's Young Leaders (QYL) with Cambridge University. He was a British Council Fellow and an Adjunct Professor, Guest Faculty, College of Forestry and Environmental Sciences, Oregon State University (OSU), Corvallis, USA. He has published books and articles on gender, women, environment and development, and has lectured extensively globally.

Pranita Bhushan Udas Research Fellow at Thompson Rivers University, Canada, is a researcher with experience in interdisciplinary water resources management. She has more than 15 years of experience working on water, gender justice and equity issues. She holds a PhD from Wageningen School of Social Sciences, Department of Environmental Sciences, Wageningen University, the Netherlands. The chapter by her in this book is based on her research working at the International Centre for Integrated Mountain Development as Gender, Water Adaptation Specialist. Her other publications are on gender, participatory development, governance and multi-stakeholder processes.

Katharine Vincent is the Director of Kulima Integrated Development Solutions (Pty) Ltd in South Africa and a Visiting Associate Professor at the University of the Witwatersrand, Johannesburg. Her research interests are in adaptation to climate change, taking a gender-sensitive perspective. She is also particularly interested in ensuring research findings are appropriately communicated to policymakers and practitioners to enable impact.

Foreword

More than a billion people live in river deltas, semi-arid lands and gla-cier-dependent basins in South Asia, hotspot regions that are among the most vulnerable to climate change. Over seven years, the Collaborative Adaptation Research Initiative in Africa and Asia (CARIAA) programme supported research to strengthen resilience in these hotspots by informing policy and practice. This work brought together more than 450 researchers across 15 countries through four consortia, with selected study areas based on geographic and social similarities – with the aim of sharing knowledge and experiences across disciplines, sectors and geographies. Drawing from that work, this book provides novel insights into how climate change is reshaping the constraints and opportunities for women and men who live in these landscapes.

Beyond funding for the four research consortia, CARIAA created collab-orative spaces to explore common themes and synergies across this diverse community of partners. Key among these collaborative spaces were efforts championed by these editors and authors to incorporate gender methods into research design, comparing changes in women's agency across Africa and Asia, and work towards the present volume which provides in-depth studies from several locations across South Asia. The authors represented here have shown themselves to be thought leaders bringing their insights to the attention of scholarly and policy audiences ranging from Adaptation Futures 2018 in South Africa to the Climate Change Adaptation Policy and Science conference in Nepal.

The first half of this volume covers conceptual framings, while the latter half covers strategies available to differently positioned women. Through-out, the authors share the lived experiences of households and communities, and advance our understanding of gender and social equity, recognizing that vulnerability comes in different forms and is not limited to women. The impacts of climate change are experienced differently by different peo-ple, based on their exposure and capacity to respond to risks. The evidence presented speaks to how the gendered nature of decision-making and access to resources creates differential capacities of women and men to adapt to climate change. Capacity to adapt is determined by gender, age, ethnicity, class and household structure. The remoteness of hotspots also matters, as people living in mountains and floodplains, semi-arid lands and river deltas face distinct risks and opportunities.

Through the richness of these pages a deeper and more nuanced understanding emerges of both vulnerabilities and agency, the risks people face and their ability to overcome them. The authors demonstrate how the impacts of climate change aggravate pre-existing socio-economic vulnerabilities. In many places, the activities that women are responsible for rely on climate, such as collecting water and biomass, or rearing cattle and planting food crops. Women have tended to remain in place and continue to deal with environmental stress while men migrate in search of income elsewhere. This volume unpacks vulnerabilities to reveal the intersections of geography and social identity, the constructions of masculinity and femininity, and the complex interplay of agro-ecology and socio-economy. The authors document how perceptions of risks are conditioned by social position and identity, and how climate pressures are interacting with longer-term changes in land use and environmental degradation.

Yet women and men are not mere victims of a changing climate. The authors identify how households and livelihoods are changing, opening up new opportunities for women to exercise control over decision-making. Climate pressures and male out-migration are increasing the workload for many women, but also expanding their responsibilities at home and within their communities. Within a changing climate, women and men are engaged in an ongoing process of renegotiating their roles. In the words of the editors, "In a substantially altered world of climate change, women and men are both faced with changes that may have no precedent, no standard practice or pattern to follow". Ultimately, the insights presented here provide a glimpse into an evolving society, one in which gendered assumptions are themselves adapting to a new reality.

BRUCE CURRIE-ALDER
Program Leader,
Collaborative Adaptation Research Initiative
in Africa and Asia (CARIAA),
International Development Research Centre

Acknowledgements

This book developed from the 15th Conference of the Indian Association of Women's Studies held in Chennai in 2017, where gender and climate change was introduced for discussion as a sub-theme. The chapters presented were published in the Women's Studies section of the *Economic and Political Weekly* (EPW 2018).[1] We are obliged to *EPW* for permitting us to publish two of the papers, the first by Divya Susan Solomon and Nitya Rao, and the second by Pranita B. Udas, Anjal Prakash and Chanda G. Goodrich, in this volume.

Shoma Choudhury of Routledge provided the support that was required for this very collaborative process. We were encouraged by her to extend the theme from India to South Asia, thus enriching this volume. It now includes chapters from Bangladesh, India, Nepal and Pakistan. We acknowledge Shloka Chauhan of Routledge and B Balambigai of SPi Global for their editorial support in bringing this book into the world.

The chapters in this collection are based on research generated through the Collaborative Adaptation Research Initiative in Asia and Africa (CAR-IAA) funded by the Canadian International Development Research Centre and the UK's Department for International Development. Over five years (2014–18) CARIAA, through four consortia working in three distinct agro-ecological regions – ASSAR (Adaptation at Scale in Semi-Arid Regions), PRISE (Pathways to Resilience in Semi-Arid Economies) working in semi-arid regions, DECCMA (Deltas, Vulnerability and Climate Change: Migration and Adaptation) in delta regions, and HI-AWARE (Himalayan Adaptation, Water and Resilience) working to improve livelihoods in glacier- and snowpack-dependent river basins – contributed to the generation of research insights on the barriers and enablers to climate change in these hotspots.

CARIAA provided an intellectual platform for all the authors to engage in a collective project of social justice. Additional funding support enabled us to develop gender analysis skills amongst a wider group of researchers. We would also like to sincerely thank IDRC for helping make this book open access. We owe special thanks to Bruce Currie-Alder, Program Leader at CARIAA, for writing the foreword. We sincerely thank Amit Mitra who went beyond simple editing, contributing his deep knowledge and passion for the subject to rethink our writings.

We would also like to acknowledge a host of people, not all of whom can be individually named, who have engaged with us throughout the process of developing this book, including colleagues and friends. The chapters included are from both established and up-and-coming researchers and reflect a mix of expertise and methodologies.

In addition, the editors individually acknowledge the support received from their respective institutions and consortia: Anjal Prakash from ICI-MOD, and Bharti Institute of Public Policy, Indian School of Business, Hyderabad, India; Asha Hans and Amrita Patel acknowledge Jadavpur University, especially Prof Sugata Hazra engaged in the DECCMA project; and Nitya Rao recognizes the support from the ASSAR team of researchers and from the School of International Development, University of East Anglia.

We dedicate this book to the women who are transforming this world by their dedication to intellectual and grassroots pursuits in the field of engendering climate change.

Note

1 Review of Women's Studies. "Gender and Climate Change". *Economic and Political Weekly,* 28 April 2018. Vol LIII No 17. Pp 35–37, 38–45, 46–54, 63–69.

1 Gender, climate change and the politics of vulnerability

An introduction

Nitya Rao, Anjal Prakash, Asha Hans, and Amrita Patel

Introduction

Climate change is transforming countries the world over, and the South Asian subcontinent is no exception. Floods, heatwaves, weak monsoons and unseasonal rains, all occurring in a relatively short timeframe, are adversely affecting millions of poor people (Ramanathan et al. 2005; Amarnath et al. 2017; Vinke et al. 2017; Gunaratna 2018, Pant et al. 2018). The nature of the risks confronting their lives and livelihoods are becoming increasingly unpredictable. Worst affected are the poor and marginalized living in 'climate hotspots', namely, coastal areas, mountain ranges, semi-arid regions and cities (De Souza et al. 2015; Ford et al. 2015; Sivakumar and Stefanski 2010). Those whose livelihoods are directly dependent on agriculture and the agrarian economy remain the most vulnerable (Farooqi et al. 2005; Gentle and Maraseni 2012). Climate-induced floods in many parts of Bangladesh, for instance, are threatening the food security of the poor (Faisal and Parveen 2004; Yu et al. 2010), while in the Indus river basin of Pakistan, the lack of information and resource access further constrains people's abilities to adapt to the changing climate (Rasul et al. 2012; Abid et al. 2015).

Even though research and policy are often framed in terms of climate change impacts alone, our starting point is the recognition that climate change aggravates pre-existing socio-economic vulnerabilities and everyday risks that the poor confront in their daily lives (Allen et al. 2018; IPCC 2019). Such 'contextual vulnerability' is based on a processual and multidimensional view of climate–society interactions (O'Brien et al. 2007). There is growing recognition in research on climate change that both perceptions of risk and its impacts on people are associated with social position and identity, in particular gender, and of the ways in which this intersects with other factors like class, caste/ethnicity and age (Dankelman et al. 2008; MacGregor 2010; Goodrich et al. 2019).

The impacts of climate change could prove particularly severe for some women, yet very little research that is both disaggregated and contextually embedded is available (Cannon 2002; Demetriades and Esplen 2008; Djoudi et al. 2016; Pearse 2017). This book is an attempt to fill this gap, focusing on the gendered experiences of climate change across different geographies and social contexts in South Asia. Moving beyond the binaries of women and men, we conceptualize gender in terms of the unequal power relations between women and men across social groups (Rao and Kelleher 2005; Rao et al. 2019). We also include the implications of these intersecting (dis)advantages for both material conditions and social positions, which in turn shape strategies for coping with and adapting to climate change (Young 2016).

In this introduction, we set out the key issues present in the debates on climate change from a gender and South Asian perspective, emphasizing the need to take on board the specificities of locational and cultural vulnerabilities, experiences and opportunities, in order to sensitively address issues of adaptation. We understand vulnerability as "a state of susceptibility to harm from exposure to stresses associated with environmental and social change and from the absence of capacity to adapt" (Adger 2006: 268; Otto et al. 2017). Vulnerability is closely associated with the multiple risks – climatic, market, social – that people confront, but equally with the social, cultural, political and institutional contexts in which they are embedded (Rao et al. 2020). Adaptation, then, includes the "plans, actions, strategies or policies to reduce the likelihood and/or consequences of risks or to respond to such consequences" (IPCC 2018).

The chapters in this collection are based on research generated through the Collaborative Adaptation Research Initiative in Asia and Africa (CARIAA) funded by the Canadian International Development Research Centre and the UK Department for International Development. Over the past five years, CARIAA has sought to understand climate change impacts on vulnerable communities through four consortia working in three distinct agro-ecological regions: ASSAR (Adaptation at Scale in Semi-Arid Regions) and PRISE (Pathways to Resilience in Semi-Arid Economies) working in semi-arid regions; DECCMA (Deltas, Vulnerability and Climate Change: Migration and Adaptation) in delta regions; and HI-AWARE (Himalayan Adaptation, Water and Resilience) working on glacier- and snowpack-dependent river basins. The authors include both established and up-and-coming researchers, and the chapters reflect a mix of expertise and methodologies. Empirical evidence is drawn from quantitative, qualitative and participatory research.

Gender and climate change: a view from South Asia

The global literature focusing on gender and climate change, though still evolving (Denton 2002; Dankelman 2010; Arora-Jonsson 2011), establishes with high confidence that gender relations are an integral part of climate-change processes and the social transformations these set in motion (Pearse 2017). This is because the immediate effects of changing climate are felt in the natural resource-dependent sectors – agriculture, forestry and water – where women's involvement in farming, water collection and use, or biomass collection for cooking, is historically sizeable (Dixon-Mueller 2013).

With the increasing unpredictability of rainfall and temperatures, men tend to migrate to urban and peri-urban areas as a pre-emptive risk-reduction strategy, leaving women behind in the rural areas to bear the near total burden of everyday survival in their absence (Rajasree 2010; Rao et al. 2019).

A view from South Asia does not appear very different on the surface, though this remains an under-researched area, given the deep fissures created by caste, marital status and class. Widows and separated women, or women from the lower castes, are often subject to subtle forms of social exclusion, in addition to confronting the effects of frequent floods and droughts, the lowering of water tables, growing scarcity of water and reduction in yields of forest biomass. The negative effects on grain yields of rainfall variability and temperature changes may also increase malnutrition and related health risks (Woodward et al. 1998). While men, not confined to the same extent by domestic and caring responsibilities, have more opportunities than women, and often migrate to earn a living, caste inequalities mediate the outcomes for them too. Poorer and lower-caste men from marginal geographies experience harsh working environments, resulting in ill-health, ultimately increasing the work and care burdens borne by women (Awumbila and Momsen 1995; Rao and Raju 2019).

Caste and class often coincide in South Asia, especially in India and Nepal. Poor households here, especially those in rural areas and dependent on farming or agricultural labour, are likely to be the worst affected, with women in such households facing significant care deficits on account of the increased workload of ensuring household reproduction and care (Ahmed and Fajber 2009; Rao and Raju 2019). Adapting to climate change, then, requires a better understanding not just of the gender divisions of labour, but equally of the mechanisms through which access to and control over natural and other resources are negotiated (Sultana 2014). This in turn implies attention to gender relations, including male contributions to adaptation, whether through migration, other productive activities or support with reproductive work; governance, especially in the provision of basic services and infrastructure; and the enhancement of basic capabilities, including appropriate knowledge, skills and technology.

Within research and policy, the word 'gender' often replaces 'women', not fully taking on board the socially constructed positions that gender relations entail. In a review article asking how far adaptation, vulnerability and resilience research has engaged with gender over the past decade, Bunce and Ford (2015) found that of the 123 articles they reviewed, only one focused on men, and none on other sexual identities. What is still evident in research, but more so in policy, is that women continue to be typically portrayed as overburdened, weak and vulnerable, rather than exercising agency in numerous different ways to cope with adversity and make a living, and more importantly, contributing to meeting their future aspirations for their children (Ghosh et al. 2018; Mitra 2018). Men are invisible, and if at all mentioned, their absence for work is seen as increasing women's burdens and vulnerabilities (Arora-Jonsson 2011; Rao et al. 2017).

What this brings to the fore is the need to understand the changing power relationships between men and women, but equally the intersecting nature of identities and how they play out in terms of enabling or preventing access to resources and services (Rao et al. 2020). Thus, rather than focusing on women as a unified social category, the impacts of climate change need to be differentiated to better understand how women within socially excluded and economically insecure groups, such as the landless, small-holders, indigenous people or lower-caste groups, may be more intensely affected than other more privileged women (Rao and Raju 2019). In the absence of a differentiated analysis, policy interventions will be unable to strengthen the adaptive capacities of those most in need. Research and policy therefore need to go beyond the collection of sex-disaggregated data to develop an understanding of the multiple, intersecting disadvantages confronting women and men of different social categories and the opportunities available to them for exercising agency in adapting to change (cf. Fraser 1989).

The South Asian region includes seven countries, four of which – Bangladesh, India, Nepal and Pakistan – are included in this volume, primarily because these were the countries where CARIAA research was conducted. While we do not have specific chapters from Sri Lanka, Maldives or Bhutan, we have tried to integrate key insights from these countries in this introductory chapter. Maldives, in particular, as a Small Island Developing State (SIDS), is extremely vulnerable to climate change, especially slow-onset hazards, such as coastal erosion, sea-level rise, salinity intrusion, and change in monsoon patterns and hence rainfall (Government of Maldives 2009). A study in the capital city of Male revealed that more than 50 per cent of respondents viewed sea-level rise as a real threat affecting the country (Stojanov et al. 2017). One of the few studies on the gendered impacts of climate change in the Maldives focuses on changes in the tuna-fishing industry. Similar to other forms of fishing, a gender division of labour exists, with women involved in post-harvest and value-addition activities (Asian Development Bank 2014). Climate change and associated disasters have impacted the production and processing of tuna (Fulu 2007). Despite the guarantee of gender equality in the constitution of the Maldives, women tend to lose out both economically and socially (El-Horr and Pande 2016).

Sri Lanka, also an island state, has some similarities to the Maldives, for instance, its proneness to disasters such as the 2004 tsunami. Based on post-tsunami interviews with 40 widows and widowers, Hyndman (2008) demonstrates that while men could consider remarrying, this was a harder option for women. Yet, this has meant a loss of social support and security. Ethnic conflict, especially in the Northern Province, already poverty-stricken and home to displaced people, and dependent primarily on agricultural livelihoods, has intensified some of these effects (Mani et al. 2018). Fishing is another industry which has been adversely affected; women have lost incomes and been forced to look for other options, often more burdensome and less rewarding. Girls' education has suffered in this process (APWLD 2011). Only very recently have some women received support to sustain their livelihoods through the Adaptation Fund (Withanachchi 2019).

Migration is a major issue in Sri Lanka, as in other South Asian countries discussed in this book. Among other aspects of gender discrimination, for instance, the Government of Sri Lanka's Family Background Report (FBR) of 2013 includes policy requiring prospective women migrant domestic workers to document their reproductive and family histories when applying to work overseas. In 2015, the FBR was revised to include women domestic workers in Sri Lanka's tea and rubber estates. The inclusion of the plantation clause raised questions about the state's intention to regulate women's reproductive capacity, alongside restricting the mobility of women wishing to work within and outside the country (Gunathilaka et al. 2018). Similar restrictions were imposed on the migration of young women in Bangladesh (Siddiqui 2003).

Bhutan, a landlocked country in the Himalayan mountains, is increasingly vulnerable to climate change, as witnessed in erratic and extreme rainfall events and glacial lake outburst floods (GLOFs) (ICIMOD 2016). Rising temperatures and uncertain precipitation have made rural settlements prone to landslides, floods and water shortages. With men increasingly moving out in search of work, without credit or alternate sources of employment, women are experiencing growing hardship in their daily lives. They are confined to the informal sector, working hard for low wages, often engaging in risky enterprises carrying a threat of HIV/AIDS, as in other South Asian contexts (ICIMOD 2016). These examples confirm that patriarchy, or the institutionalization of male dominance, manifests itself in different ways across the subcontinent, and to understand these complexities, one needs to consider both social and class positions that mediate the normative boundaries of gender relations across these societies (Hyndman 2008).

In this volume, we focus on a few selected themes that demonstrate diverse strategies and mechanisms for coping and adaptation to climate variability and change. These include male migration, the formation of women's collectives, shifts to drought- and flood-resilient cash crops, and changing gender roles, among others. Contexts of vulnerability and pressures for survival, while intensifying hardship, potentially create spaces for the creative use of agency, and facilitate the loosening of rigid social norms, in the process contributing also to changes in gender and wider social relations. We find, then, an ongoing process of negotiation and renegotiation of roles and responsibilities, but equally decision-making and voice within the household. In its interface with some form of collective action for overcoming technology and scale constraints, social norms in the domains of both production and reproduction are challenged. How far these processes will contribute to attaining both gender justice and climate justice, however, remains to be seen.

Contributions to the volume

At the 15th Conference of the Indian Association of Women's Studies in 2017 held in Chennai, for the first time the theme of gender and climate change in India was introduced and the papers presented were later published in the Women's Studies section of the *Economic and Political Weekly (EPW)* (2018).[1] Together with two of the papers published in *EPW*, this book goes beyond the Indian situation to open up the debate on the realities of climate change and its gendered impacts across the subcontinent. The wider focus on the South Asian countries of India, Pakistan, Bangladesh and Nepal has brought in nuances of gender relations across borders, presenting the many similarities and differences across locational, socio-cultural and policy contexts. The chapters explore the links between climate variability and environmental change, along with other political economy factors, the precariousness of livelihoods and women's work burdens both in production and reproduction, and their health and nutritional wellbeing and indeed very survival, links that have hardly been explored in the case of South Asia.

Having set out the conceptual framings and assumptions underpinning the study of gender and climate change in South Asia in this introductory chapter, the book goes on to use empirical cases from South Asia to explore the effects of increasing climate variability on poor women and men – their social roles, agency and indeed adaptive capacities. In particular, it focuses on the notions of contextual and relational vulnerabilities (Taylor 2013; Turner 2016), examining how issues such as water scarcity or drought are addressed across contexts, pointing in the process to the gendered nature of water access (Yadav and Lal 2018) and the changing gender roles therein (Kulkarni 2018). We also explore a range of adaptive options available to differently positioned women (Bhatta and Aggarwal 2016; Rao et al. 2019). By adaptive options, we mean "the array of strategies and measures that are available and appropriate for addressing adaptation. They include a wide range of actions that can be categorized as structural, institutional, ecological or behavioural" (IPCC 2018: 542). While male migration emerges as an important adaptation strategy (Bhatta et al. 2016; Jha et al. 2018; Ahsan 2019), the chapters examine the implications of a range of activities that differently positioned women undertake including changing cropping patterns, livestock rearing and petty business to reduce their everyday vulnerabilities. Experiences of women and men in response to alterations in the social-ecological system, be they real or perceived climate stressors, have implications for inter- and intra-household relations and in turn on the wellbeing of individuals (Sarkar 2017; Goodrich et al. 2019).

Unpacking vulnerabilities: intersections of geography and social identity

While the concept of vulnerability is not a new one, emerging from early literature on famines and the failure of entitlements (Sen 1981) which recognized the political, economic and social embeddedness of people's experiences and responses to crises, there is relatively less evidence on the gendered nature of vulnerabilities, especially in response to climatic shifts. With growing acknowledgement of its importance for adaptation planning, a small body of work has been emerging, largely from sub-Saharan Africa. In the case of South Asia, the lack of disaggregated and nuanced information makes writing on this theme a challenge.

Further, much of the literature on climate change in South Asia continues to be framed in complex scientific and technical language, making it difficult to evoke a more fluid sense of climate change as a process affecting the lived experiences of people, women and men, in different parts of the region (cf. O'Brien et al. 2007). It is thus through an intellectually collaborative process that we have tried to use critical gender analytical tools to enable a more nuanced understanding of the implications of climate change for gender relations and household wellbeing. An attempt is made to unpack the gendered nature of risks and vulnerabilities confronting people's lives and livelihoods, the relations of power and social complexities in terms of intersections of gender, caste and class that shape their choices, and the trade-offs between individual interests and reciprocal exchanges, in adapting to changing circumstances (Rao 2017; Rao et al. 2019). Intersectionality, then, is at the core of conducting a gender analysis of the impacts of climate change (see Qaisrani and Batool in this volume).

Based on the different vulnerabilities and experiences of women across our study locations, we draw attention to the complexities of analyzing climate change. In South Asia, changing climate, whether shrinking glaciers and reduced snow coverage, cyclones, flooding and land erosion, or increasing aridity due to declining rainfall, is an additional stressor, intensifying the burdens of existing poverty. In this volume, Abbasi et al. analyse the role of culture and geography in shaping gendered vulnerabilities and coping with climate change in Pakistan. Climate change in the mountain ecosystems of Nepal and India have affected the water regimes in these areas, in turn altering gender roles and responsibilities, the ownership, rights and control over resources, as well the very constructions of masculinity and femininity in these contexts (see Tamang and Udas in this volume).

Solomon and Rao, while exploring the implications of depleting ground water levels in rural Tamil Nadu for household livelihoods, emphasize the ways in which the masculine connotations of borewell technology, and the state and community discouragement of women entering these spaces, differentially impact women in this location. The authors observe a simultaneous reversal of the gendered binary in relation to borewells at the household level as women's assets, such as gold, are being called upon for deepening wells or digging new ones, making them partners in these assets. Yet, the loss of control over their gold and its investment in borewells is impacting women's marriage choices, decision-making and overall wellbeing. Udas, Prakash and Goodrich in their chapter from the flood- and drought-prone West Champaran district of Bihar, India, also find women's gold and dowries being increasingly used in response to climate stress and extreme poverty, but given the unequal power relations in this location, the women are unable to secure adequate social returns for this contribution. With dowry carrying connotations of son preference, households with more daughters or only daughters have become the most vulnerable in the locality. Both these chapters demonstrate the importance of social location and gendered ideologies in relation to asset control in shaping women's capacities to respond to stressors.

Apart from institutions of the state and society, the effects of climate are also interlinked across scales. Chandni Singh explores the implications of changing patterns of water availability, access and use in a water-scarce district of Karnataka, India, focusing particularly on gendered vulnerability at different scales: the household and intra-household level; settlement and community level; and larger, socio-ecological scale. She notes a shift from a culture of water harvesting and storage to one of unmitigated water extraction, leading to shortages of drinking water. With uneven purchasing power, not all households, especially women-headed households, are able to buy water due both to the high costs involved and lack of time or transport to travel to towns for this purpose. Families unable to buy water then consume unpotable water with severe health implications, and this is the case for a majority of households headed by divorced or widowed women in the study area.

Adaptive strategies: agricultural diversification, migration and collective action

In the substantially altered world of climate change, women and men both face unprecedented changes, with no standard practices to follow. Livelihood shifts brought about by climate variability, whether in terms of the choice of crops or diversification into non-farm employment, often involve migration from the home. With reduced labour availability, households including those headed by women, in the absence of their men, often turn to cash crops such as sugarcane or vegetables, which are resistant to climate stressors, but also less labour intensive (Udas et al., this volume). This however can result in over-extraction of soil nutrients and water, and while bringing in incomes, could negatively affect the quality of diets, nutrition and health. In this volume, we understand adaptation as the "the process of adjustment to actual or expected climate and its effects, in order to moderate harm or exploit beneficial opportunities" (IPCC 2019:542). From a gender perspective, this would entail integrating the needs of poor and marginalized women, but equally ensuring that they are not further marginalized during the adaptation process (Wigand et al. 2017; Kaufmann 2019; Rao et al. 2019).

Migration as adaptation

Migration from the hotspots is often viewed as a distress response, with people leaving home as their geographical remoteness limits their access to new livelihood opportunities. Migration can, however, have contradictory effects on both livelihoods and gender relations. In some situations, climate change-linked migration can lead to improving economic situations (Hazra et al. and Vincent et al., this volume). At the same time, there are parallel trajectories of exploitation and autonomy accompanying male migration. While women are left in charge of the home and everyday decision-making, their limited mobility and constrained access to public places and institutions accentuates their vulnerability and limits their adaptive capacities. Yet there is a difference between left-behind wives receiving remittances from their husbands, even occasionally, and female-headed households with no male support. Based on their study in Bangladesh, Vincent et al. conclude that fewer women in non-migrant households considered themselves happy compared to those living in migrant households. This is driven partly by women's marital status, but also by the nature of migration itself, whether planned and supported, or forced.

The complexity of migration is visible in Nepal (Tamang et al., this volume), where men more than women have taken advantage of new market opportunities. Carrying the burden of agricultural work, women labour on average nearly 3.5 hours more than men every day. While much of this work involves drudgery, it interestingly also challenges traditional gender divisions of labour, with women taking over work designated as male, such as ploughing and hoeing (Rao et al. 2020). This added responsibility includes new roles for women such as hiring labour, selling livestock and engaging with off-farm activities, usually male prerogatives. Despite higher work and care burdens in often adverse working conditions, it appears that women's fiscal agency is strengthened by changes in power relations within households with increased assets resulting from male migration (see Hazra et al., this volume).

Migration of women as an adaptation strategy is a new area of research and policy advocacy. The paper by Abbasi et al. (this volume) sets the scene for a better understanding of the processes involved, pointing also to the differences that exist within groups of women, as seen in the case of women in up-stream, mid-stream and down-stream areas of the river Indus in Pakistan. The subtle differences across agro-ecologies and social groups highlight the contextual nature of gender relations and roles, and the need for disaggregated perspectives in order to respond effectively to local needs and priorities. Yet as mentioned earlier, legal and policy frameworks, as in the cases of Sri Lanka and Bangladesh, can either support or impede the opportunities available to different categories of women within these countries. The adaptation consequences of migration, then, depend not just on the nature of migration itself, but also on the gendered nature of social institutions that either facilitate or block equitable opportunities for women and men across contexts.

Collective action and resilience

Climate change and the resulting livelihood diversification, including male outmigration, has opened new spaces for women's agency, individual and collective. While some scholars focus on the expansion of women's roles within households (Hazra et al. and Singh in this volume), further research is required to explore the expansion of agency this may entail. Collective action has been the subject of a little more research, as in the Indian Bengal Delta, where women have established spaces for mutual support through self-help groups (Ghosh et al. 2018). In some instances, these groups have facilitated opportunities for women to seek social and economic stability, and avoid risky ventures including sex work. Such organizational space itself opens up a multitude of opportunities and networks, going beyond specific activities linked to microcredit (Kalpana 2017). In West Champaran, in addition to sugarcane cultivation, women left behind have turned to shared animal care as an adaptation strategy as in their perception it is less hazardous than agriculture, while also ensuring mutual support (Udas, Prakash and Goodrich this volume).

Women's significant roles as agents of change in promoting climate-change mitigation action needs to be recognized and supported both individually and collectively (see Qaisrani and Batool, this volume). Rural women's entrepreneurial potential provides one of the entry points enabling women to become change leaders, but this involves overcoming strong cultural constraints to women's mobility and public participation, especially in rural settings. These papers all highlight the link between the two and the importance of collective action for overcoming extreme forms of vulnerability, including challenging entrenched forms of power and inequality.

Conclusion

Environmental and climatic change is a reality affecting the poor and resourceless the worst. Without a proper understanding of the complexity of issues involved, the links between the geophysical, the agro-ecological and the socio-economic, we will be unable to suggest gender-sensitive policies and support mechanisms. While the risks and vulnerabilities resulting from climate change which intensify gender inequalities have been emphasized in the case of South Asia, our fine-grained analysis points to the enhancement of women's agency in these locations and suggests strategies for supporting such agency. Ownership of assets remains a major constraint to gender equality and can be overcome by new adaptive strategies especially built on women's collectives. Similarly, in contexts of male migration, women's decision-making powers are supplemented by a loosening of gender roles in work, yet they need to be supported in these tasks by social and technological innovations that can ensure that their effort and drudgery does not increase (Rao et al. 2019).

Gender roles and responsibilities are a social construct and vary across communities and geographies. By combining these variables, we have sought to achieve a deeper understanding of the complexity of climate-change impacts. While our empirical evidence and analysis is located in South Asia, the articles collectively demonstrate that women, especially in marginal, rural households, have common interests not just in transforming the external environment affected critically by climate change but also in redefining their lives and homes. With the increasing unpredictability of agriculture due to climate change and past and present investments, both public and private, that have undermined the sustainable use of resources, they seek greater cooperation and reciprocity from their husbands and wider kin networks to ensure a life of security and dignity. Drawing specifically from the concrete realities of women's lives, we can construct a larger theoretical argument around women's critical contributions to household livelihoods, more so in an era of climate change. Examining the changing nature of gender relations, roles and responsibilities could help revisit the debates around the "patriarchal bargain" (Kandiyoti 1988).

The chapters in this book seek to deepen and nuance our understanding of gender relations in the context of climate change. They also foreground the need for further research that can identify and highlight innovative adaptation mechanisms on the ground, often led by poor women with little support from state or community institutions. Only by giving recognition to women's agency and their contributions to both paid and unpaid work, to the production and reproduction of their households, can we create an enabling environment which is sensitive to their needs and gives them a voice in decision-making mechanisms. It is not just the issue of vulnerability vs agency, but a better understanding of changing gender relations within households and communities, that will help develop and support sustainable and equitable livelihoods and in turn adaptation.

Note

1 Review of Women's Studies. 'Gender and Climate Change', *Economic and Political Weekly*, April 28, 2018. Vol LIII No 17. Pp. 35–37, 38–45, 46–54, 63–69.

References

Abid, M.E.A., Scheffran, J., Schneider, U.A., and Ashfaq, M. 2015. 'Farmers' perceptions of and adaptation strategies to climate change and their determinants: The case of Punjab province, Pakistan', *Earth System Dynamics*, 6(1): 225–243.

Adger, W.N. 2006. 'Vulnerability', *Global Environmental Change*, 16(3): 268–281.

Ahmed, S., and Fajber, E. 2009. 'Engendering adaptation to climate variability in Gujarat, India', *Gender & Development*, 17(1): 33–50.

Ahsan, R. 2019. 'Climate-induced migration: Impacts on social structures and justice in Bangladesh', *South Asia Research*, 0262728019842968.

Allen, M.R., Dube, O.P., Solecki, W., Aragón-Durand, F., Cramer, W., Humphreys, S., Kainuma, M., Kala, J., Mahowald, N., Mulugetta, Y., Perez, R., Wairiu, M., and Zickfeld, K. 2018. 'Framing and Context' In: *Global Warming of 1.5°C. An IPCC Special Report on the impacts of global warming of 1.5°C above pre-industrial levels and related global greenhouse gas emission pathways, in the context of strengthening the global response to the threat of climate change, sustainable development, and efforts to eradicate poverty* [Masson-Delmotte, V., P. Zhai, H.-O. Pörtner, D. Roberts, J. Skea, P.R. Shukla, A. Pirani, W. Moufouma-Okia, C.

Péan, R. Pidcock, S. Connors, J.B.R. Matthews, Y. Chen, X. Zhou, M.I. Gomis, E. Lonnoy, T. Maycock, M. Tignor, and T. Waterfield (eds)]. In press

Amarnath, G., Alahacoon, N., Smakhtin, V., and Aggarwal, P. 2017. *Mapping multiple climate-related hazards in South Asia* (Vol. 170). International Water Management Institute (IWMI).

APWLD. 2011. *Climate justice briefs: Rural women's adaptation strategies.* Asia Pacific Forum on Women, Law and Development. Sri Lanka.

Arora-Jonsson, S. 2011. 'Virtue and vulnerability: Discourses on women, gender and climate change', *Global Environmental Change*, 21: 744–751.

Asian Development Bank. 2014. *Maldives: Gender equality diagnostic of selected sectors.* Mandaluyong City: Asian Development Bank.

Awumbila, M., and Momsen, J.H. 1995. 'Gender and the environment: Women's time use as a measure of environmental change', *Global Environmental Change*, 5(4): 337–346.

Bhatta, G.D., and Aggarwal, P.K. 2016. 'Coping with weather adversity and adaptation to climatic variability: a cross-country study of smallholder farmers in South Asia', *Climate and Development*, 8(2): 145–157.

Bhatta, G.D., Aggarwal, P.K., Poudel, S., and Belgrave, D.A. 2016. 'Climate-induced migration in South Asia: Migration decisions and the gender dimensions of adverse climatic events', *Journal of Rural and Community Development*, 10(4).

Bunce, A., and Ford, J. 2015. How is adaptation, resilience and vulnerability research engaging with gender? *Environmental Research Letters*, 10: 123003. http://iop-science.iop.org/article/10.1088/1748-9326/10/12/123003

Cannon, T. 2002. 'Gender and climate hazards in Bangladesh', *Gender & Development*, 10(2): 45–50.

Dankelman, I., Alam, K., Ahmed, W.B., Gueye, Y.D., Fatema, N., and Mensah-Kutin, R. 2008. *Gender, climate change and human security: Lessons from Bangladesh, Ghana and Senegal.* New York: WEDO, ABANTU for Development, ActionAid and ENDA.

Dankelman, I. 2010. *Gender and climate change: An introduction.* London: Earthscan.

De Souza, K., Kituyi, E., Harvey, B., Leone, M., Murali, K.S., and Ford, J.D. 2015. 'Vulnerability to climate change in three hot spots in Africa and Asia: KEY issues for policy-relevant adaptation and resilience-building research', *Regional Environmental Change*, 15: 747. doi:10.1007/s10113-015-0755-8

Demetriades, J., and Esplen, E. 2008. 'The gender dimensions of poverty and climate change adaptation', *Ids Bulletin*, 39(4): 24–31.

Denton, F. 2002. 'Climate change vulnerability, impacts, and adaptation: Why does gender matter?', *Gender & Development*, 10(2): 10–20.

Dixon-Mueller, R.B. 2013. *Rural women at work: Strategies for development in South Asia.* New York: RFF Press.

Djoudi, H., Locatelli, B., Vaast, C., Asher, K., Brockhaus, M. and Sijapati, B.B. 2016. 'Beyond dichotomies: Gender and intersecting inequalities in climate change studies', *Ambio*, 45(3): 248–262.

El-Horr, J., and Pande, R.P. 2016. *Understanding gender in Maldives: Towards inclusive development (English).* Direction in development. Country and regions. Washington, DC: World Bank Group. http://documents.worldbank.org/curated/en/448231467991952542/Understanding-gender-in-Maldives-towards-inclusive-development

Faisal, I.M., and Parveen, S. 2004. 'Food security in the face of climate change, population growth, and resource constraints: Implications for Bangladesh', *Environmental Management*, 34(4): 487–498.

Farooqi, A.B., Khan, A.H., and Mir, H. 2005. 'Climate change perspective in Pakistan', *Pakistan Journal of Meteorology*, 2(3).

Ford, J.D., Berrang-Ford, L., Bunce, A., McKay, C., Irwin, M., and Pearce, T. 2015. 'The status of climate change adaptation in Africa and Asia', *Regional Environmental Change*, 15(5): 801–814.

Fraser, N. 1989. *Unruly practices: Power, discourse and gender in contemporary social theory*. University of Minnesota Press.

Fulu, E. 2007. 'Gender, vulnerability, and the experts: Responding to the Maldives tsunami', *Development and Change*, 38(5): 843–864.

Gentle, P., and Maraseni, T.N. 2012. 'Climate change, poverty and livelihoods: Adaptation practices by rural mountain communities in Nepal', *Environmental Science & Policy*, 21: 24–34.

Ghosh, A.K., Banerjee, S., and Naaz, F. 2018. 'Adapting to Climate Change – induced Migration', *Review of Women's Studies, Economic and Political Weekly* 53 (17): 63–69.

Goodrich, C.G., Udas, P.B., and Larrington-Spencer, H. 2019. 'Conceptualizing gendered vulnerability to climate change in the Hindu Kush Himalaya: Contextual conditions and drivers of change', *Environmental Development*, 31: 9–18. https://doi.org/10.1016/j.envdev.2018.11.003

Government of Maldives. 2009. National Adaptation to Climate Change. A background paper prepared by the Ministry of Housing, Transport and Environment for the Maldives Partnership Forum (MPF) to be held in Maldives, 23–24 March 2009:1

Gunaratna, K.L. 2018. 'Managing climate change in South Asia.' In *Towards equitable progress* (pp. 53–69). Singapore: Springer.

Gunathilaka, R.P.D., Smart, J.C.R., Fleming, C.M., and Hasan, S. 2018. 'The impact of climate change on labour demand in the plantation sector: The case of tea production in Sri Lanka', *Australian Journal of Agricultural and Resource Economics*, 62(3): 480–500. doi:10.1111/1467-8489.12262

Hyndman, J. 2008. 'Feminism, conflict and disasters in post-tsunami Sri Lanka', *Gender, Technology and Development*, 12(1):101–121.

ICIMOD. 2016. Bhutan Climate + Change Handbook by Bhutan Media and Communications Institute Bhutan and International Centre for Integrated Mountain Development Kathmandu. http://lib.icimod.org/record/32399/files/icimodBhutanClimate016.pdf

IPCC. 2018. 'Annex I: Glossary' [Matthews, J.B.R. (ed.)]. In: *Global Warming of 1.5°C. An IPCC Special Report on the impacts of global warming of 1.5°C above pre-industrial levels and related global greenhouse gas emission pathways, in the context of strengthening the global response to the threat of climate change, sustainable development, and efforts to eradicate poverty* [Masson-Delmotte, V., P. Zhai, H.-O. Pörtner, D. Roberts, J. Skea, P.R. Shukla, A. Pirani, W. Moufouma-Okia, C. Péan, R. Pidcock, S. Connors, J.B.R. Matthews, Y. Chen, X. Zhou, M.I. Gomis, E. Lonnoy, T. Maycock, M. Tignor, and T. Waterfield (eds)]. In Press.

IPCC. 2019. 'Summary for Policymakers.' In *IPCC Special Report on the ocean and cryosphere in a changing climate* [H.-O. Pörtner, D.C. Roberts, V. Masson-Delmotte,

14   *Nitya Rao et al.*

P. Zhai, M. Tignor, E. Poloczanska, K. Mintenbeck, A. Alegría, M. Nicolai, A. Okem, J. Petzold, B. Rama, N.M. Weyer (eds)]. In press.

Jha, C.K., Gupta, V., Chattopadhyay, U., and Amarayil Sreeraman, B. 2018. 'Migration as adaptation strategy to cope with climate change: A study of farmers' migration in rural India', *International Journal of Climate Change Strategies and Management*, 10(1): 121–141.

Kalpana, K. 2017. *Women, microfinance and the state in neo-liberal India*. New Delhi: Routledge.

Kandiyoti, D. 1988. 'Bargaining with patriarchy', *Gender and Society*, 2 (3): 274–290.

Kaufmann, W. 2019. 'Vulnerabilities in a Wetter World: A study on migration as an adaptation strategy to climate change, with under-five mortality as an intermediating variable.' Department of Economics, Uppsala University, Sweden. Accessed on 12 Nov 2020 from http://urn.kb.se/resolve?urn=urn:nbn:se:uu:diva-376870.

Kulkarni, S. 2018. 'Gender, water and well-being.' In Narain, V., Goodrich, C.G., Chourey, J., and A. Prakash (eds). *Globalization of water governance in South Asia*, pp. 19–34. New Delhi: Taylor & Francis, Routledge.

MacGregor, S. 2010. 'Gender and climate change: From impacts to discourses', *Journal of the Indian Ocean Region*, 6(2): 223–238. doi:10.1080/19480881. 2010.536669.

Mani, M., Bandyopadhyay, S., Chonabayashi, S., Markandya, A., and Mosier, T. 2018. *South Asia's hotspots: The impact of temperature and precipitation changes on living standards*. Washington, DC: World Bank. 10.1596/978-1-4648-1155-5.

Mitra, A. 2018. 'Male migrants and women farmers in Gorakhpur', *Economic and Political Weekly*, 53(17): 55–62.

O'Brien, K., Eriksen, S., Nygaard, L.P., and Schjolden, A. 2007. 'Why different interpretations of vulnerability matter in climate change discourses', *Climate Policy*, 7: 73–88.

Otto, I.M., Reckien, D., Reyer, C.P., Marcus, R., Le Masson, V., Jones, L., Norton, A. and Serdeczny, O. 2017. Social vulnerability to climate change: A review of concepts and evidence. *Regional Environmental Change*, 17(6):1651–1662.

Pant, G.B., Kumar, P.P., Revadekar, J.V., and Singh, N. 2018. 'Climate and climate change: An overview.' In *Climate change in the Himalayas*, pp. 1–38. Cham: Springer.

Pearse, R. 2017. 'Gender and climate change', *Wiley Interdisciplinary Reviews: Climate Change*, 8(2): e451.

Rajasree, R. 2010. 'Gender and climate change', *Commodity Vision*, 3(5): 99–101.

Ramanathan, V., Chung, C., Kim, D., Bettge, T., Buja, L., Kiehl, J.T., Washington, W.M., Fu, Q., Sikka, D.R., and Wild, M. 2005. 'Atmospheric brown clouds: Impacts on South Asian climate and hydrological cycle', *Proceedings of the National Academy of Sciences*, 102(15): 5326–5333.

Rao, A. and Kelleher, D., 2005. Is there life after gender mainstreaming? *Gender & Development*, 13(2), 57–69.

Rao, N., and Raju, S. 2019. 'Gendered time, seasonality, and nutrition: Insights from two Indian Districts', *Feminist Economics*, 26(2): 95–125.

Rao, N. 2017. 'Assets, agency and legitimacy: Towards a relational understanding of gender equality policy and practice', *World Development*, 95(3): 45–54. doi:10.1016/j.worlddev.2017.02.018.

Rao, N, Lawson, E. T., Raditloaneng, N, Solomon, D and Angula, M. 2017. 'Gendered Vulnerabilities to Climate Change: Insights from the Semi-Arid

Regions of Africa and Asia,' *Climate and Development* 11(1): 14–26. doi: 10.1080/17565529.2017.1372266.

Rao, N., Mishra, A., Prakash, A., Singh, C., Qaisrani, A., Poonacha, P., Vincent, K. and Bedelian, C. 2019. 'A qualitative comparative analysis of women's agency and adaptive capacity in climate change hotspots in Asia and Africa', *Nature Climate Change*, 9: 964–971.

Rao, N., Singh, C., Solomon, D., Camfield, L., Sidiki, R., Angula, M., Poonacha, P., Sidibe, A., and Lawson, E.T. 2020. 'Managing risk, changing aspirations and household dynamics: Implications for wellbeing and adaptation in semi-arid Africa and India', *World Development*, 125.

Rasul, G., Mahmood, A., Sadiq, A., and Khan, S.I. 2012. 'Vulnerability of the Indus delta to climate change in Pakistan', *Pakistan Journal of Meteorology*, 8(16).

Sarkar, R. 2017. *Women and water: Adaptation practices to climate variability in the Sikkim Himalaya* (Doctoral dissertation).

Sen, A. 1981. 'Ingredients of famine analysis: Availability and entitlements', *The Quarterly Journal of Economics*, 96(3): 433–464.

Siddiqui, T. 2003. 'An Anatomy of Forced and Voluntary Migration from Bangladesh: A Gendered Perspective.' In: Morokvasic M., Erel U., Shinozaki K. (eds) *Crossing Borders and Shifting Boundaries*. Schriftenreihe der Internationalen Frauenuniversität »Technik und Kultur«, 10: 155–176. Wiesbaden: VS Verlag für Sozialwissenschaften.

Sivakumar, M.V., and Stefanski, R. 2010. 'Climate change in South Asia.' In *Climate change and food security in South Asia*. Dordrecht: Springer.

Stojanov, R., Duží, B., Kelman, I., Němec, D., and Procházka, D. 2017. 'Local perceptions of climate change impacts and migration patterns in Malé, Maldives', *The Geographical Journal*, 183(4): 370–385.

Sultana, F. 2014. 'Gendering climate change: Geographical insights', *The Professional Geographer*, 66(3): 372–381.

Taylor, M. 2013. 'Climate change, relational vulnerability and human security: Rethinking sustainable adaptation in agrarian environments', *Climate and Development*, 5(4): 318–327.

Turner, M.D. 2016. 'Climate vulnerability as a relational concept', *Geoforum*, 68: 29–38.

Vinke, K., Martin, M.A., Adams, S., Baarsch, F., Bondeau, A., Coumou, D., and Robinson, A. 2017. 'Climatic risks and impacts in South Asia: Extremes of water scarcity and excess', *Regional Environmental Change*, 17(6): 1569–1583.

Wigand, C., Ardito, T., Chaffee, C., Ferguson, W., Paton, S., Raposa, K., Vandemoer, C. and Watson, E. 2017. 'A climate change adaptation strategy for management of coastal marsh systems', *Estuaries and Coasts*, 40(3): 682–693.

Withanachchi, A. 2019. 'Despite all odds: climate resilient women in Sri Lanka.' https://www.adaptation-undp.org/despite-all-odds-climate-resilient-women-sri-lanka

Woodward, A., Hales, S., and Weinstein, P. 1998. 'Climate change and human health in the Asia Pacific region: Who will be most vulnerable?', *Climate Research*, 11(1): 31–38.

Yadav, S.S., and Lal, R. 2018. 'Vulnerability of women to climate change in arid and semi-arid regions: The case of India and South Asia', *Journal of Arid Environments*, 149, 4–17.

Young, Z.P. 2016. 'Gender and development.' In *Handbook on gender in world politics*. Cheltenham: Edward Elgar Publishing.

Yu, W., Alam, M., Hassan, A., Khan, A.S., Ruane, A., Rosenzweig, C., Major, D. and Thurlow, J. 2010. *Climate change risks and food security in Bangladesh*. London: Routledge.

Part I
Vulnerabilities

2 Vulnerabilities of rural women to climate extremes

A case of semi-arid districts in Pakistan

Ayesha Qaisrani and Samavia Batool

Introduction

Gender is often discussed in the discourse on climate change, yet rarely forms a core element of research and policy perspectives. Recent literature recognizes that climate change impacts are not gender neutral but affect women disproportionately (Babugura 2010; MacGregor 2010; Rao et al. 2017). Vulnerabilities associated with climate change are accentuated because of women's pre-existing marginalization and may lead to 'inequity traps'. Conversely, climate adaptation options are also gendered and may not always result in equal increased resilience for women, girls, men and boys (UNDP 2013; Angula and Menjono 2014). Such growing recognition of disparate climate change impacts calls for more gender-just development and climate policies so that women's adaptive capacities can be enhanced, irrespective of their varying socio-economic status (Tiani et al. 2016).

Given the current focus in the literature on differential climate impacts (UNDP 2013; Rao et al. 2017; Petesch et al. 2018), it is pertinent to consider the geographical aspects that may exacerbate existing climate change threats for both men and women. Accordingly, this chapter focuses on semi-arid locales and assesses how men and women are affected by climate change in a water-scarce geographical setting.

This chapter does not homogenize the experiences of rural women in the context of climate change, rather it illustrates the various pathways through which women's interaction with the environment and natural resources shapes their vulnerability. Following Hahn et al.'s (2009) and Panthi et al.'s (2016) conceptualizations of vulnerability, women's vulnerability is analyzed in terms of their exposure, sensitivity and adaptive capacity to climate change impacts. It is recognized that women's vulnerability to climate change not only depends on their gender; rather, other factors such as their age, geographic location, socio-economic background, household structure and position in the household and community play a role in defining their exposure, sensitivity and adaptive capacity to climate change. In that sense, the chapter brings into play discussions on intersectionality, with context-specific examples taken from the fieldwork that help explain women's vulnerability in different situations (Nizami and Ali 2017). Furthermore, we briefly touch upon the climate change policy framework of Pakistan and provide a policy-oriented way forward for enhancing the resilience of vulnerable women in rural semi-arid regions.

The insights emerge from two PRISE projects in Pakistan focusing on migration and cotton value chains in the semi-arid regions of Dera Ghazi Khan, Faisalabad and Mardan districts.[2,3] For this study, we focus on rural areas in Dera Ghazi Khan and Faisalabad districts,[4] Punjab province. Based on mixed-methods data collected between December 2016 and January 2017, this chapter draws on household surveys (n = 200 per district) and gender-segregated focus group discussions (n = 8 per district, 4 each with men and women). Two union councils (UCs) were targeted in each district, and one village was selected from each union council. UC Sokar (village: Mor Jhangi) and UC Kala (village: Basti Raman) were selected in Dera Ghazi Khan and UC 116 (village: Bhainian) and UC 39 (village: Gangapur) were selected in Faisalabad. The household survey was targeted at mainly farming households, most of whom were cotton farmers. Table 2.1 summarizes the sample selection from each district with respect to landholding.

In addition, we carried out focus group discussions with between eight and twelve women and men of different economic backgrounds in the age group 18–65 years. The participants were randomly selected from the stratified sample list of the households surveyed. The discussions in the focus groups were targeted at gathering information about the gendered impacts of climate change, gender roles and responsibilities in post-disaster scenarios, and perceptions about gender differences in climate vulnerability. All the focus group participants were either full- or part-time workers in cotton fields. Women were mostly involved in cotton picking, whereas men were responsible for cultivation and crop management.

This chapter is organized as follows. We begin by introducing the contextual background of the study areas, then present the field evidence, and finally summarize our findings and offer some conclusions.

Contextual background

Pakistan, a country of over 208 million people as of 2017 (Population Census 2017), has a varied geographical terrain ranging from the coastal areas in the south to the mountainous ranges of Hindu Kush, Karakoram and the Himalayas in the north. The country lies in the arid-humid ecological range, with 88 per cent of its area characterized as arid to semi-arid (Farooq et al. 2009). According to the Pakistan Meteorological Department (PMD) maps, about 25–30 per cent of Pakistan's total area receives rainfall of 250–500mm annually and is classified as semi-arid (Rasul et al. 2012). These semi-arid regions contain some of the country's most fertile land and serve as the breadbasket for the growing population (Saeed et al. 2016). These areas are densely populated and natural resource-based livelihoods are the mainstay of the rural people. For instance, Dera Ghazi Khan has a population density of 245.7/km^2 and Faisalabad has a population density of 1321/km^2 (Population Census 2017).

Table 2.1 Sample selection by district

District	Union Council	Village	No. of landless agrarian workers	No. of small landholders (owning less than 12.5 acres of land)	No. of large landholders (owning land greater than 12.5 acres)	No. of non-farm households (non-agrarian)	Total
Dera Ghazi Khan	Kala	Basti Raman	25	25	25	25	100
	Sokar	Morjhangi	25	25	25	25	100
Faisalabad	UC 116	Bhainian	25	25	25	25	100
	UC 39	Gangapur	25	25	25	25	100

Source: Qaisrani et al. (2018).

In terms of future climate risks for Pakistan, the Intergovernmental Panel on Climate Change (IPCC) Fifth Assessment Report (AR5) finds a higher likelihood of increase in warming, glacial melting rate and precipitation rate (CDKN 2014). The annual mean temperature alone has increased by 0.57°C over the last century. Not only this, heatwave days increased by 31 from 1981 to 2007 (ADB 2017). Similarly, high variations are found in the precipitation pattern. An increase in average monsoon precipitation, for example, was observed in Punjab region during 2009–2015, which triggered heavy flooding in 2010, 2011 and 2012 in different parts of Punjab (ibid). These conditions are impacting soil quality and water availability (IPCC 2013). As a result, increased desertification and land degradation are shrinking arable lands, leading to declining agricultural production, complete crop failures through floods or pest attacks, lower incomes, health issues for livestock and humans through the spread of water- and heat-borne disease, and adverse impacts on lives and property during extreme events (Hussain 2010). Unreliable rains expose the agriculture sector to serious jeopardy – too much rain, especially in the northern regions, lead to floods downstream, washing away standing crops, while a lack of rain creates a drought-like situation in the semi-arid plains, hampering agricultural production (Zaman et al. 2009; Hasson et al. 2017). Unsustainable crop production threatens to further worsen the food insecurity situation in Pakistan, as declining natural resources have to feed a rapidly growing population (Rasul et al. 2012; Nizami and Ali 2017).

Limited availability of water particularly in semi-arid lands makes these regions highly vulnerable to the changing climate. Future projections of water availability for semi-arid regions show a downward trend towards the end of this century (Bates et al. 2008). It means that semi-arid regions are on the verge of convergence with arid and hyper-arid lands (ibid.). In addition, demand for water for agricultural purposes in arid and semi-arid regions of Asia is also projected to rise by 10 per cent in the case of a 1°C temperature increase (Fischer et al. 2002; Liu 2002).

Both Dera Ghazi Khan and Faisalabad are in the wheat- and cotton-producing belt of semi-arid Punjab (Batool and Saeed 2017). About 80 per cent of the population of Dera Ghazi Khan is rural (Pakistan Bureau of Statistics 2017), largely agriculturists growing wheat, cotton, sugar cane and rice. The Dera Ghazi Khan district contains the Koh-e-Suleiman mountains on the eastern border, while its western side is marked by sandy plains and the Indus River (Malana and Khosa 2011). Some of the villages near the plains are irrigated by extensive canal systems; however, many located closer to the mountain ranges are irrigated by unreliable mountain streams (ibid.) which often flood the farms during monsoon rains, causing havoc for the poor farmers (Salik et al. 2017). Saeed et al. (2014) report that over the five years between 2009 and 2014, the frequency of rainfall decreased in South Punjab, where Dera Ghazi Khan is situated; however, the intensity of the monsoon rains increased. For the rest of the year, the climate is dry, the rainfall is scanty and the district suffers from water shortage, especially in areas not irrigated by the canal system (Malana and Khosa 2011). The district historically struggles with waterlogging and salinity issues due to seepage from canals built during the colonial period (LEAD Pakistan 2015). This makes the groundwater bitter and unfit for drinking. Many villages depend on the government to supply drinking water, which is provided through pipes at certain filter points (Qaisrani et al. 2018). Table 2.2 shows access to safe drinking water and sanitation facilities in the study districts.

As noted above, even though average annual rainfall levels have not changed much, the timings and patterns of rainfall have changed, with consequences for natural resource-based livelihoods (WFP 2018). In particular, the late arrival of monsoon rains disrupts the crop growing season, adversely affecting farmers' yields (ibid.). In addition, the frequency of excessive rains has increased over time, causing flooding in the area (Salik et al. 2017; WFP 2018). These factors, along with low levels of socio-economic

Table 2.2 Socio-economic indicators by district

S. no	Indicator	Dera Ghazi Khan	Faisalabad	Year and Source
1.	Population	2,872,201	7,873,910	2017 (Population Census Summary)
2.	Percentage of rural population	80%	52%	2017 (Population Census Summary)
3.	Net primary enrolment rate for women (rural)	51%	72%	2014–15 PSLM
4.	Net primary enrolment rate for men (rural)	71%	72%	2014–15, PSLM
5.	Women's literacy (15 years and older) (rural)	16%	48%	2014–15, PSLM
6.	Men's literacy (15 years and older) (rural)	47%	66%	2014–15, PSLM
7.	Improved drinking water facility (tap water and motor pump) (rural)	37%	70%	2014–15, PSLM
8.	Improved sanitation facility (rural)	45%	86%	2014–15, PSLM
9.	Multidimensional poverty rate (district level)	63.7%	19.4%	GoP, 2016
10.	Male to female sex ratio (district level)	108.2	108.6	Population Census, 2017

Source: Multiple sources.

development, have gendered implications. Men are deemed responsible for crop production, crop protection, sale of output and water management. Women contribute at various stages of crop production; however, their farm work largely goes unrecognized (Samee et al. 2015). They are also actively involved in livestock management. Dera Ghazi Khan has one of the highest multidimensional poverty rates in Pakistan, with 63.7 per cent of its population living in poverty (GoP 2016). Such wide-ranging poverty and inequality in access to services and opportunities results in a lower socio-economic status for many women in the district.

On the other hand, while the land in the rural areas of Faisalabad is very fertile, it is also the largest industrial district of Pakistan. The urban–rural ratio is fairly even in the district with about 52 per cent of its population being rural (Pakistan Bureau of Statistics 2017). The rural areas are irrigated through the extensive canal system. Maintenance problems, however, cause waterlogging and salinity. Groundwater is brackish and unsuitable for human consumption or even agricultural or livestock usage due additionally to industrial waste polluting the water quality (D'Souza 2006; Salik et al. 2017).

The maximum summer temperature in Faisalabad can go up to 50°C, while the winter lows can drop down to -2° (Abbas 2013). The average temperature of the district is increasing, especially in the winter months (Cheema et al. 2006). Cheema et al. (2006) report an increase of 0.22°C in Faisalabad between 1945 and 2004. Erratic rainfall and a rise in the frequency and intensity of heat waves are increasingly affecting crop production in the area (IPCC 2013; Saeed et al. 2016). Mueller et al. (2014) report that heat waves in winter in Faisalabad adversely affect wheat yields. Average temperature projections from KNMI Climate Explorer (2017) show increasing temperature trends for both Faisalabad and Dera Ghazi Khan (Figures 2.1 and 2.2).

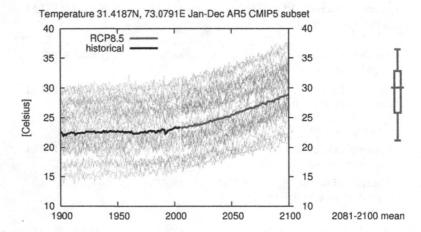

Figure 2.1 Temperature trend for Faisalabad.

Source: KNMI Climate Explorer with RCP 8.5 (2017).

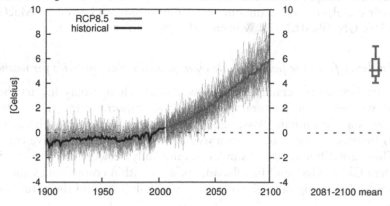

Figure 2.2 Temperature trend for Dera Ghazi Khan.

Source: KNMI Climate Explorer with RCP 8.5 (2017).

Due to a thriving industrial sector, the socio-economic status of Faisalabad is better than that of Dera Ghazi Khan. The multidimensional poverty rate of Faisalabad is 19.4 per cent, significantly lower than in Dera Ghazi Khan (GoP and UNDP 2016). Net primary enrolment rates for both men and women are the same at 72 per cent, however, adult female literacy is much lower (48 per cent) than adult male literacy (66 per cent) (Pakistan Bureau of Statistics 2015a, b) (Table 2.2). The labour force participation rate for women (27.3 per cent) is higher than the national average (23.7 per cent); however, female participation is about three times lower than male labour force participation (Khan et al. 2014; World Bank 2018). In the absence of recent data, 2008 estimates show that 60.5 per cent of the female labour force was employed in agriculture, 21.4 per cent in industry and 18.1 per cent in services in Faisalabad (Khan et al. 2014).

While socio-economic indicators reflect better conditions for gender equality in Faisalabad (cf. Table 2.2), traditional gender roles and patriarchal structures reflect similar trends to those observed for Dera Ghazi Khan (Khan et al. 2013; Akhtar et al. 2018). A report by the Pakistan Commission for the Status of Women (PCSW 2018) shows that in rural Punjab, married women in the age group 15–64 spend an average of 29 hours a week on unpaid domestic work. Rural women from poorer economic backgrounds in both Dera Ghazi Khan and Faisalabad are often involved in selling milk, eggs and meat, which allows them to earn a meagre income. Despite their active involvement in farming, most women are scarcely involved in decision making regarding land use, livestock sale and purchase, and crop choices (Ishaq and Memon 2016; PCSW 2018).

Against this background, this study attempts to assess how climate change may influence gender relations in Dera Ghazi Khan and Faisalabad. In particular, it explores the different ways in which climate change is perceived, experienced and responded to by different individuals based on their social, cultural, economic and demographic backgrounds (MacGregor 2010; UNDP 2013; UN Women 2018; Petesch et al. 2018).

Evidence from the field: climate change and vulnerabilities for women

A contextual understanding of how climate change may impact the lives of women in agricultural households is important for the analysis of how socio-economic class, age and generation interact with each other to increase or decrease women's vulnerabilities and adaptative capacities. There are differences and similarities not only between the two districts of Dera Ghazi Khan and Faisalabad, but also within communities and households, depending on their roles and responsibilities and the nature of their work.

Class matters

Women of diverse socio-economic classes experience climate change differently. In both Dera Ghazi Khan and Faisalabad, women's experience of climate extremes is quite different between those belonging to landless or small-farm households and those belonging to better-off households with large landholdings. About 63 per cent of private farms in Punjab are 5 acres or smaller, and cover only 22 per cent of total land in Punjab, revealing that a majority of the rural population are smallholders (Agricultural Census 2010). Poorer women from households with no land or from smallholdings are more involved in agricultural activities than women from large landholder households and better socio-economic backgrounds (UN Women 2018). They are usually involved in sowing, weeding, crop picking and harvesting, while men are largely responsible for land preparation, irrigation and harvesting (Samee et al. 2015). As well as working on their own farms, women from poorer households are hired as daily wage labourers for cotton picking during the harvest season, especially in Dera Ghazi Khan (Batool and Saeed 2018). Their direct involvement in and dependence on agriculture makes their livelihoods more at risk, subjecting them to income shocks if agricultural productivity is affected by climate extremes. As rural areas are transforming at different rates, unlike Dera Ghazi Khan, where poorer women are still largely engaged in agricultural activities, in Faisalabad, they are more likely to engage with non-farm activities such as paid domestic work (Salik et al. 2017; Qaisrani et al. 2018). Such non-agricultural jobs safeguard women from experiencing the direct impacts of climate change and hence lessen their vulnerability.

In an earlier paper (Qaisrani et al. 2018), we showed that Faisalabad in general had lower scores on livelihood vulnerability, owing to higher adaptive capacity than Dera Ghazi Khan and lesser exposure to climate change impacts. Dera Ghazi Khan appears to be experiencing the double jeopardy of higher exposure to climate change in terms of frequent floods plus dry spells and rising temperatures, contributing to a higher incidence of complete crop failure. With high levels of multidimensional poverty, a high dependency ratio and lower access to climate information, the population has lower adaptive capacities to respond to these impacts (ibid.). Poorer women told us how difficult it was to work in the fields, especially in the summer, when temperatures rise as high as 49°C (PMD 2018). They reported that they usually get sunburnt and often experience heatstroke. This risk pattern is particularly visible in young girls and elderly women. Sajida, 18, from a landless agrarian household in village MorJhangi in Dera Ghazi Khan, said: "It is getting difficult for us to pick cotton due to extreme heat in the field these days. My grandmother got ill the other day due to heat and she had to rest for a week."

Unreliable rains and frequent floods in Dera Ghazi Khan adversely impact women's work patterns (Zaman et al. 2009; Hasson et al. 2017). They often have to spend more time and energy on crop maintenance and livestock management, leaving them with little time to spare for domestic responsibilities. According to secondary district-level data, only 37 per cent of households in rural Dera Ghazi Khan have access to improved drinking water facilities and a mere 45 per cent have improved sanitation facilities (MICS 2018) (Table 2.2). Women from about 26 per cent of households have to walk more than 30 minutes to access drinking water (MICS 2018). With more time spent on fetching water, the time for vital household activities like cooking and child care is reduced.

Whether they are working in their own fields or working as day-wage labourers, women's efforts, although crucial, are valued less in monetary terms. Many women farmers work for minimal remuneration (much lower than men's) in other farmers' fields. Their work on the family farm goes unrecognized socially as they are considered 'helpers' of their male counterparts (cf. Rao et al. 2019). For waged workers, the low remuneration is further adversely affected during shocks like floods. Women's incomes decline further and some even lose their jobs when crops fail. Salma, 37, a cotton picker on a large landholder's farm in village MorJhangi, Dera Ghazi Khan said: "Men can easily migrate for work whereas we have to stay here (at home) to take care of the family. After the floods (of 2010), my daily wage was decreased from Rs. 200 ($1.62) per bale of cotton (1 bale = 177.8 kg) to Rs. 75 ($0.61). Even before the floods, my husband earned more than me (Rs. 350 per bale = $3.2)."

This quote reveals how men and women experience vulnerability differently in case of climate-related events. It not only reflects the gendered wage disparities between men and women but also highlights the differences in opportunities available to them. While the livelihoods of both men and women are affected by floods, men are more likely to find other opportunities by migrating. Women are less likely to do so owing to both mobility restrictions and their domestic responsibilities. This sentiment was also observed in Faisalabad, where cotton picking is one of the main livelihood activities for poorer women.

In both districts, men are regarded as the primary breadwinners and women as 'helpers' and minor contributors to the household income. In village Gangapur in Faisalabad, 35-year-old Asma Bibi, who belongs to a small-landholder household, said: "Men are usually over-burdened as they have to work hard and earn for all of us (household members). They have to go out of town to find work whereas we (women) just have to stay at home or we just go to the fields nearby to pick cotton, which is not a difficult task."

While such sentiments show that women undermine their own contribution to agricultural work, they also reflect the neglect that women in agriculture have historically experienced in these areas (Samee et al. 2015). In addition, working in the fields was considered necessary for women to make ends meet in the lower-income households, and not a sign of empowerment. As 29-year-old Naveeda, belonging to a landless household in Basti Raman in Dera Ghazi Khan remarked: "I work on their (large landowner's) field to earn some extra money during harvest season, but women's place is at home to take care of their family".

In Dera Ghazi Khan and Faisalabad, women are rarely seen as the principal farmers. Discussions in both districts showed that women are often deemed 'secondary farmers' due to their 'seasonal' or 'need-based' involvement in particular stages of agricultural production. In cotton production, their involvement is particularly important at the time of picking. They are contracted by large farm-holders at lower wages than men. However, their income from this activity is often the only remuneration they receive in the year for their own work. In wheat production as well, in both the districts, lower-income women were involved in wheat harvesting but were largely paid in kind rather than in cash. The women do not consider cotton picking too draining physically but harvesting wheat is perceived to be a difficult task and often the entire family, including young girls, participate in the process. In Faisalabad, it was reported that scores of young girls take extended breaks from school to help their families pick cotton. Humeira Khalid, 36, a school teacher in village Bhainian in Faisalabad said: "One of the best students of my class has left school and is not appearing in exams because it's the harvest season. She is working on the farm."

Such physically exhausting work takes its toll on the health of the women working in the fields. Rising temperatures and lack of rains increase the risk of pest attacks on crops (Memon 2011). As a result, farmers increase the use of pesticides, which can be injurious to health due to lack of proper safety measures for cotton picking. In both Dera Ghazi Khan and Faisalabad, cotton pickers, mostly women and young girls, risk developing skin allergies from pesticide residues on the cotton bolls. These risks are particularly high for pregnant cotton pickers (Batool and Saeed 2017).

Women who are not directly involved in farm activities are also vulnerable to secondary impacts of climate events. The secondary impacts often manifest themselves in reduced household income, which has significant implications for food security and spending on health and education. Not only this, women are now also pursuing more off-farm employment to supplement household income while still maintaining existing household responsibilities. In Dera Ghazi Khan, for instance, many women from lower- and middle-income backgrounds who stay at home contribute to the household income by embroidering scarves and shawls for private contractors. However, the remuneration is nominal and the task quite time-consuming, as noted also in other studies of women's home-based work such as the lace makers of Narsapur (Mies 1982). Tehmina, 27, from Dera Ghazi Khan said: "One *dupatta* (shawl) takes around two weeks to complete. We have to look after the household too, and this is slow work. At the end of the month, we only get PKR 100–150 ($1.25) for a dupatta that the contractors sell for thousands in the market. They earn profits on our hard work."

Women from lower-income households are generally more mobile within the vicinity of the village. They are often also involved in water collection and doing domestic chores for other households. In Faisalabad, women's mobility appeared to be more relaxed than in Dera Ghazi Khan across income groups. Most women from lower-income households in Faisalabad reported that they are free to move within the village to collect water or to do paid domestic work. Women from both lower- and middle-income groups said they had the freedom to visit relatives and friends. Women here were more visible in public spaces such as markets and streets, walking around without being accompanied more often than in Dera Ghazi Khan. This difference between the two districts is also captured by the Women's Economic and Social Empowerment (WESE) Index of 2018, according to which the score for women in rural Faisalabad is 54 (medium category) compared to Dera Ghazi Khan, where the score is 34 (low category) (UN Women 2018).

Among the wealthier households or the more politically powerful families in the villages, women's mobility is restricted due to notions of the families' 'honour' and 'prestige'. In gender-segregated focus group discussions in Dera Ghazi Khan, many respondents, both women and men, felt that 'noble' women are expected to stay at home, while the men are responsible for earning. Religious interpretations and social norms structure gender roles in society, where women's responsibility is to look after the private realm, while the men's duty is to earn and provide for the household. Seema, a 22-year-old woman in Basti Raman in Dera Ghazi Khan, who was married at 16 to a politically powerful man in the village, took the view that she does not need to go out to work when she can live like a 'princess' in her home.

At the community level, affluent households are less likely to be vulnerable to climate impacts, due to relatively higher adaptive capacity (Batool and Saeed 2018). Women from wealthier households in both the districts are less likely to engage in climate-sensitive activities like farm labour or be involved in collecting water. For educated women from the economically better-off households, teaching is considered the most respectable occupation, if they work at all. Healthcare work is considered acceptable by women from middle-income backgrounds; however, for women from economically or politically powerful backgrounds in the village (such as the village chief or large landowners), healthcare work involving door-to-door services is not socially acceptable. Men and women from households across income and landholding categories in both areas preferred women to be involved in home-based occupations such as tailoring or embroidery, although this was often not possible for the poor.

Inter-generational dynamics and decision making

Interactions with women and men farmers in Dera Ghazi Khan and Faisalabad showed that generational differences also shape gendered climate vulnerability. About half of rural households in the study sites included large extended families, which confer varying roles, status and power on women according to their social and economic position. The intra-household dynamics in joint families are much more complex than in nuclear ones. The position of a woman in such a household is defined according to her relation to the head of the household as well as other women in the family. Household dynamics also determine women's role in decision making and access to resources, defining their level of vulnerability (Nizami and Ali 2017). The older women in the family usually have the highest decision-making powers and are often consulted by the men in major household matters, including household resource management and migration decisions. These trends were observed across the two districts, reflecting a broader social culture at play in the province.

For example, in response to the declining profitability of agriculture, Zahida (aged 37) recalls that her husband left MorJhangi (Dera Ghazi Khan district) and decided to seek work outside the village. Her mother-in-law was involved in the decision as to where her husband should move to, but being a new bride, she was not involved in the discussion. She mentioned that she used to face mobility restrictions imposed by her mother-in-law, who did not allow her to go out of the house alone, even to run errands, but as her children are grown up now, these restrictions have eased.

Men's migration may also shift the household dynamics regarding responsibilities and decision making. We thought that with men migrating, women might have greater economic and social status and decision-making authority. However, due to the wide prevalence of joint or extended family systems, especially in Dera Ghazi Khan, the migrant's brother often takes over as the head of the household. Sometimes, as in the case of Zahida, women's mobility is further restricted to avoid gossip about them while their husbands are away. However, these changes in decision-making authority also depend on the age and household status of women. Older women may enjoy more authority in household affairs, but younger women, especially newly married ones, may face further restrictions in their husband's absence.

While there are considerable differences between households, young girls from poorer families are more likely to be taken out of school if the household income falls due to a climate shock. Even for better-educated women, work opportunities in the village are scanty, thus their educational qualifications often do not match their livelihoods or earning potential. For example, Nasima, 57, belonging to a large landholder household in MorJhangi (district Dera Ghazi Khan), complained that while she had ensured that her daughter completed her bachelor's degree, the girl is now unemployed. There are no opportunities in the village and her father does not allow her to move to the city to work. Nasima feels that the investment in her daughter's education has been wasted.

Besides inter-generational dynamics, decision-making responsibilities within the household are often starkly gendered. While women are active in farming activities in both the study areas, their contribution to farm-related decision making, across socio-economic categories, is limited. As mentioned earlier, many women consider their roles as 'fixed' as opposed to socially constructed and do not think their contribution to agriculture and as the family caregiver is underrated. This further reinforces the exclusion of women from decision-making processes in agriculture and climate-related matters at the household level and undermines their ability to deal with climate shocks. Most of the women farm-labour respondents in Dera Ghazi Khan reported that they were not consulted about any farm-related decisions after the mega-floods of 2010. Men, on the contrary, said that they took most of the decisions in consultation with women household members, especially the older ones. Despite these consultations, however, men usually have the final say in most matters. Shabbir, a 62-year-old small landowner in Basti Raman in Dera Ghazi Khan stated: "We discuss at home what we should grow on the farm between wheat and cotton. My wife wants to grow vegetables, but I also have to see what would be more profitable in the market."

Day-to-day farm-level decisions regarding crops, seeds, fertilizer and pesticide choices and the area under each crop are often taken independently by men. Men farmers in Dera Ghazi Khan and Faisalabad felt they should be the ones taking farm-level decisions as women lack technical education about crop management, including checking insect population levels, soil moisture and crop water requirements. Here, the difference in the level of education between men and women farmers plays a major role. Most male farmers interviewed had completed primary school while their wives were mostly uneducated or informally educated, leading men to assume that women have limited skills and information. This stereotype prevails in the local administration, evident from the fact that agricultural training and information about new technologies are directed towards men. Shabana, a 57-year old widow in Faisalabad, manages her own small market garden. She receives money from her son working in Karachi, but she complains that she could make money by selling her produce if she had training on how to increase production. She informed us that she sometimes has to rely on her brother-in-law for agricultural advice.

The exclusion of women from agricultural extension training also results in information closure, with women's access to critical information limited by system design. In line with this, men reported a higher degree of awareness about climate change and its impacts on agricultural productivity than women. The majority of the women respondents in both areas were unaware of any current or potential impacts of changes in weather patterns on the crop. In the villages in both Dera Ghazi Khan and Faisalabad, the major factor limiting women's access to information was the absence of any rural women's organization to facilitate information dissemination and ensure women's participation in agricultural decision making. While some non-governmental organizations are working in rural Dera Ghazi Khan to empower women by enhancing their skills, they largely provide training in tailoring and embroidery. Men, on the other hand, are represented by a number of formal and informal small-farmers' organizations which are also a major source of climate risk information at community level. Discussions with women in Dera Ghazi Khan revealed that women did not feel the need for women farmers' organizations. This perhaps reflects the deep-rooted patriarchal structures that limit women's capacity to think of themselves as involved in farm-level decision making.

Similarly, a dichotomy exists between men and women about perceived needs related to adaptation. Most men in both the study regions felt that women have no role in adaptation decision making as they are not fully aware of climate impacts and lack technical knowledge about farm management. Women are excluded from day-to-day farm-related decisions; thus, they too feel that they cannot contribute effectively to farm management decisions before and after climate disasters. Hence, social and cultural factors play a major role in compounding the adverse impacts of climate extremes for women in rural Pakistan.

It is worth noting here that while adaptation strategies related to productive processes (e.g., earning an income, maintaining agricultural activity) are more male-centric, women play an important role in private adaptation in response to shocks and stresses (Angula and Menjono 2014). Evidence from the field surveys in Dera Ghazi Khan and Faisalabad shows that in response to food insecurity due to climate shocks, men largely think in terms of modification of farm production, diversification of income activities, migration and relying on external help as effective coping strategies. This approach clearly reflects their control over assets and social mobility. Women, on the other hand, rely on short-term solutions like reducing food consumption, selling off household assets and reducing non-food expenditure to deal with food insecurity.

Conclusion

The accounts presented in this chapter reflect the fact that how women are affected by extreme climate events depends on the context and their involvement in agriculture and natural-resource management. Women in poorer socio-economic regions, such as Dera Ghazi Khan, tend to be more vulnerable to adverse climate situations than those in slightly better-off areas, such as Faisalabad. However, not all women in a specific region are affected uniformly. We have argued that within these regions, gender differentiation in experiences and responses to extreme climate events are not only class-based but also depend on women's intra-household position. Contextual factors such as the nature of men's and women's work, their control over productive resources, decision-making power and access to opportunities for learning, determine their experiences in relation to climate change. Our results reflect that normatively, while cultural norms in both the districts are largely similar, they are less rigid in Faisalabad, giving women greater mobility and the chance of engaging in non-climate-sensitive employment opportunities, making them less vulnerable than women in Dera Ghazi Khan.

However, such opportunities are often limited for the majority of women from poorer socio-economic backgrounds, who labour in the fields of the big landowners. Women from slightly higher socio-economic classes prefer home-based work, like making handicraft items, over agricultural labour, despite the extremely low returns, as it allows them to earn staying at home. Women from more educated and better-off backgrounds prefer to be employed in the healthcare or teaching sectors. A higher socio-economic position and lesser involvement in climate-sensitive activities decreases their vulnerability.

Our study shows that while older women have higher involvement in household decision making, men still dominate crucial adaptation decisions, especially those related to agriculture. This is not only the result of a patriarchal culture of control over assets and resources, but of patriarchal government policies that target only men for agricultural training and extension programmes. Although the National Climate Change Policy (2012) and the Framework for Implementation of National Climate Change Policy (2014) recognize the need for mainstreaming gender concerns in adaptation measures, all the action points are expressed in gender-neutral language and do not specifically recognize the need to enhance the adaptive capacity of rural women. While the socially constructed gender roles do put women at risk of higher vulnerability to climate change, recognizing women and men's "differentiated but complementary" roles (Rao et al. 2017) within the household and society may allow policy makers to better address the needs of the rural population, and poor women in particular. In that sense, building climate resilience for both women and men cannot just be driven by mainstreaming gender in climate policy: it also requires women's needs to be integrated into all other national and local policies and actions that shape and construct discriminatory gender roles in society and within households. This approach, in fact, shows that vulnerability is not an inherent characteristic of women; rather it is generated by the existing discrimination that women experience (Morchain et al. 2015). Understanding the contextual factors is important for determining women's level of vulnerability to climate change, in order to devise mechanisms to improve their resilience.

Notes

1 Pathways to Resilience in Semi-arid Economies (PRISE) is a five-year, multi-country research programme that aims to create new knowledge about climate-resilient and equitable economic development in semi-arid regions of Asia and Africa.
2 For more information, please refer to Salik et al. (2017) and Qaisrani et al. (2018).
3 For more information, please refer to Batool and Saeed (2017) and Batool and Saeed (2018).
4 Districts are the second-order administrative units of Pakistan, after provinces.

References

Akhtar, Saira, Ramzan, Shazia, Ahmad, Shabbir, Huifang, Wu, Imran, Shakeel and Yousaf, Haroon. 2018. 'Women in Agriculture – Lack of Access to Resources (An Analytical Study of District Faisalabad, Punjab, Pakistan)', *SSRG International Journal of Economics Management Studies*, 5(10): 16–24.

Angula, Margaret and Menjono, Ewaldine. 2014. 'Gender, Culture and Climate Change in Rural Namibia', *Journal for Studies in Humanities and Social Sciences*, 3(1&2): 225–238.

Asian Development Bank (ADB). 2017. *Climate Change Profile of Pakistan*. Asian Development Bank (ADB): Manila, Philippines.

Babugura, Agnes. 2010. *Gender and Climate Change: South Africa Case Study*. Cape Town: Heinrich Böll Stiftung–Southern Africa.

Bates, Bryson, Kundzewicz, Zbigniew, Wu, Shaohong and Palutikof, Jean. (eds). 2008. *Climate change and water*. Technical Paper of the Intergovernmental Panel on Climate Change, IPCC Secretariat, Geneva.

Batool, Samavia and Saeed, Fahad. 2017. *Mapping the Cotton Value Chain in Pakistan: A Preliminary Assessment for Identification of Climate Vulnerabilities and Pathways to Adaptation*. Islamabad: Pathways to Resilience in Semi-arid Economies (PRISE). https://prise.odi.org/research/mapping-the-cotton-value-chain-in-pakistan-a-preliminary-assessment-for-climate-vulnerabilities-and-pathways-to-adaptation/. Accessed on 15 October 2017.

Batool, Samavia and Saeed, Fahad. 2018. *Towards a Climate Resilient Cotton Value Chain in Pakistan: Understanding Key Risks, Vulnerabilities and Adaptive Capacities*. Islamabad: Pathways to Resilience in Semi-arid Economies (PRISE). https://prise.odi.org/research/towards-a-climate-resilient-cotton-value-chain-in-pakistan-understanding-key-risks-vulnerabilities-and-adaptive-capacities/. Accessed on 14 October 2017.

Cheema, Muhammad Asghar, Farooq, Muhammad, Ahmad, Rashid, and Munir, Hassan. 2006. 'Climatic trends in Faisalabab (Pakistan) over the last 60 years (1945–2004)', *Journal of Agriculture & Social Sciences*, 2(1): 42–45.

Climate and Development Knowledge Network (CDKN). 2014. *The IPCC's Fifth Assessment Report: What's in it for South Asia?* https://cdkn.org/wp-content/uploads/2014/04/CDKN-IPCC-Whats-in-it-for-South-Asia.pdf. Accessed on 17 August 2018.

D'Souza, Rohan. 2006. 'Water in British India: The Making of a Colonial Hydrology', *History Compass*, 4(4): 621–628.

Farooq, Umar, Ahmad, Munir and Saeed, Ikram. 2009. 'Enhancing Livestock Productivity in the Desert Ecologies of Pakistan: Setting the Development Priorities', *The Pakistan Development Review*, 48(4): 795–820.

Fischer, Gunther, Shah, Mahendra and van Velthuizen, Harrij. 2002. *Climate Change and Agricultural Vulnerability*. Laxenburg: International Institute for Applied Systems Analysis.

Government of Pakistan. 2010. *Agricultural Census: Punjab Province Tabulation*. Islamabad: Government of Pakistan, Statistics Division.

Government of Pakistan. 2012. *National Climate Change Policy*. Islamabad: Ministry of Climate Change.

Government of Pakistan. 2014. *Framework for Implementation of National Climate Change Policy*. Islamabad: Ministry of Climate Change.

Government of Pakistan. 2016. *Pakistan Economic Survey 2015-16*. Islamabad: Government of Pakistan.

Government of Pakistan. 2017. *Population Census: Provisional Summary Results of 6th Population and Housing Census*. Islamabad: Pakistan Bureau of Statistics.

Government of Pakistan (GoP) and United Nations Development Programme (UNDP). 2016. *Multidimensional Poverty in Pakistan*. Islamabad: Government of Pakistan.

Hahn, Micah, Riederer, Anne and Foster, Stanley. 2009. 'The Livelihood Vulnerability Index: A Pragmatic Approach to Assessing Risks from Climate Variability and Change—A Case Study in Mozambique', *Global Environmental Change*, 19(9): 74–88.

Hasson, Shabeh, Bohner, Jurgen and Lucarini, Valerie. 2017. 'Prevailing Climatic Trends and Runoff Responses from Hindukush-Karakoram-Himalaya, Upper Indus Basin', *Earth System Dynamics*, 8: 337–355

Hussain, Syed Sajidin. 2010. *Food Security and Climate Change Assessment in Pakistan*. Oxfam Novib: Special Program on Food Security and Climate Change.

IPCC. 2013. 'Climate Change 2013: The Physical Science Basis.' Contribution of Working Group I to Stocker, T. F., D. Qin, G.-K. Plattner, M. Tignor, S. K. Allen, J. Boschung, A. Nauels, Y. Xia, V. Bex and P. M. Midgley (eds) *Fifth Assessment Report of the Intergovernmental Panel on Climate Change*. Cambridge, United Kingdom and New York, NY, USA: Cambridge University Press, p. 1535.

Ishaq, Wajiha and Memon, Shafique Qadir. 2016. 'Roles of Women in Agriculture: A Case Study of Rural Lahore, Pakistan', *Journal of Rural Development and Agriculture*, 1(1): 1–11.

Khan, Muhammad, Akhtar, Saima, Mahmood, Hafiz Zahid and Mahmood, Kashif. 2013. 'Analysing Skills, Education and Wages inFaisalabad: Implications for Labour Market', *Procedia Economics and Finance*, 5: 423–432.

Khan, Muhammad, Mahmood, Hafiz Zahid, Akhtar, Saima and Mahmood, Kashif. 2014. 'Understanding Employment Situation of Women: A District Level Analysis', *International Journal of Gender and Women's Studies*, 2(2):167–175.

KNMI (Koninklijk Nederlands Meteorologisch Instituut) (Royal Netherlands Meteorological Institute). 2017. *Climate Explorer: Annual Temperature Data*. https://climexp.knmi.nl/selectstation.cgi. Accessed on 14 May 2017.

LEAD Pakistan. 2015. *LEAD Update: Local Adaptation Plan of Action for Dera Ghazi Khan*. Islamabad: LEAD Pakistan. http://www.lead.org.pk/lead/Publications/394-LAPA-%20Community-based%20Initiative%20to%20Reduce%20the%20Effects%20of%20Waterlogging%20&%20Salinity.pdf. Accessed on 25 July 2019.

Liu, Chunzhen. 2002. 'Suggestion on Water Resources in China Corresponding with Global Climate Change', *China Water Resources*, 2: 36–37.

MacGregor, Sherilyn. 2010. 'Gender and Climate Change: From Impacts to Discourses', *Journal of the Indian Ocean Region*, 6(2): 223–238.

Malana, Muhammad Aaslam and Khosa, Muhammad Arshad. 2011. 'Groundwater Pollution with Special Focus on Arsenic, Dera Ghazi Khan, Pakistan', *Journal of Saudi Chemical Society*, 15(1): 39–47.

Memon, Naseer. 2011. *Climate Change and Natural Disasters in Pakistan*. Karachi: Strengthening Participatory Organization (SPO). http://pnc.iucnp.org/wp/wp-content/uploads/2011/10/ClimateChange_nmemon.pdf. Accessed on 16 August 2018.

MICS (Multiple Indicator Cluster Survey). 2018. *Punjab: Survey Findings Report*. Lahore: Bureau of Statistics, Planning and Development Board, Government of Punjab.

Mies, Maria. 1982. *Lace makers of Narsapur: Indian Housewives Produce for the World Market*. London: Zed Books.

Morchain, Daniel, Prati, Giorgia, Kelsey, Frances and Ravon, Lauren. 2015. 'What if Gender Became an Essential Standard Element of Vulnerability Assessments?', *Gender and Development*, 23(3): 481–496.

Mueller, Valerie, Gray, Clark and Kosec, Katrina. 2014. 'Heat Stress Increases Long-term Human Migration in Rural Pakistan', *Nature Climate Change*, 4: 182–185.

Nizami, Arjumand and Ali, Jawad. 2017. 'Climate Change and Women's Place-Based Vulnerabilities – A Case Study from Pakistani Highlands', *Climate and Development*, 9(7): 662–670.

Pakistan Bureau of Statistics (PBS). 2015a. *Labour Force Survey 2014–15*. Islamabad: Government of Pakistan.

Pakistan Bureau of Statistics (PBS). 2015b. *Pakistan Social and Living Standard Measurement 2014–15*. Islamabad: Government of Pakistan.

Pakistan Meteorological Department (PMD). 2018. *Dera Ghazi Khan: Cities Weekly Weather Outlook*. Islamabad: PMD. http://www.pmd.gov.pk/meteorogram/punjab.php?district=Dera+Ghazi+khan&division=Dera%20Ghazi%20Khan. Accessed on 17 May 2017.

Panthi, Jeeban, Aryal, Suman, Dahal, Pyush, Bhandari, Parashuram, Krakauer, Nir, Pandey, Vishnu Prasad. 2016. 'Livelihood vulnerability approach to assessing climate change impacts on mixed agro-livestock smallholders around the Gandaki River Basin in Nepal', *Regional Environmental Change*, 16(4): 1121–1132.

PCSW (Punjab Commission on Status of Women). 2018. *Women's Economic and Social Wellbeing Survey in Punjab 2017–18*. Lahore: Government of Punjab.

Petesch, Patti, Bullock, Renee, Feldman, Shelley, Badstue, Lone, Rietveld, Annie, Bauchspies, Wenda, Kamazni, Adelbertus, Tegbaru, Amare and Yila, Jummai. 2018. 'Local Normative Climate Shaping Agency and Agricultural Livelihoods in Sub-Saharan Africa', *Journal of Gender, Agriculture and Food Security*, 3(1): 108–130.

Qaisrani, Ayesha, Umar, Muhammad Awais, Siyal, Ghamze Ali and Salik, Kashif Majeed. 2018. *Rural Livelihood Vulnerability in Semi-Arid Pakistan: Scope of Migration as an Adaptation Strategy*. Islamabad: Pathways to Resilience in Semi-arid Economies (PRISE).

Rao, Nitya, Lawson, Elaine, Raditloaneng, Wapula, Solomon, Divya and Angula, Margaret. 2017. 'Gendered Vulnerabilities to Climate Change: Insights from the Semi-Arid Regions of Africa And Asia', *Climate and Development*, 11(1): 14–26.

Rao, Nitya, Gazdar, Haris, Chanchani, Devanshi and Marium Ibrahim. 2019. 'Women's Agricultural Work and Nutrition in South Asia: From Pathways to a Cross-disciplinary, Grounded Analytical Framework', *Food Policy*, 82: 50–62.

Rasul, Ghulam, Zahid, Maida and Bukhari, Syed Ahsan Ali. 2012. *Climate Change in Pakistan: Focused on Sindh Province.* Islamabad: Pakistan Meteorological Department.

Saeed, Fahad, Suleri, Abid Qaiyum and Salik, Kashif Majeed. 2014. *Planning for Floods: Now or Never.* Policy Brief No. 44. Islamabad: Sustainable Development Policy Institute (SDPI).

Saeed, Fahad, Salik, Kashif Majeed and Ishfaq, Sadia. 2016. *Climate Change and Heat Waves: Rural to Urban Migration in Pakistan. A Silent Looming Crisis.* Islamabad: Sustainable Development Policy Institute (SDPI).

Salik, Kashif Majeed, Qaisrani, Ayesha, Umar, Muhammad Awais and Siyal, Ghamze Ali. 2017. *Migration Futures in Asia and Africa: Economic Opportunities and Distributional Effects – The Case of Pakistan.* Islamabad: Pathways to Resilience in Semi-arid Economies (PRISE).

Samee, Durre, Nosheen, Farhana, Khan, Haq Nawaz, Khowaja, Imdad Ali, Jamali, Khalida, Paracha, Pervex Iqbal, Akhtar, Shahnaz, Batool, Zahira and Khanum, Zohra. 2015. *Women in Agriculture in Pakistan.* Islamabad: Food and Agriculture Organisation of the United Nations.

Tiani, Anne Marie, Bele, Mekou Youssoufa, Kankeu, Richard Soufo, Chia, Eugene Loh and Teran, Alba Saray Perez. 2016. 'Gender and Forest Decentralisation in Cameroon: What Challenges for Adaptive Capacity to Climate Change?' In Colfer, Carol J. Pierce, Basnett, Bimbika Sijapati and Elias, Marlene (eds). *Gender and Forests: Climate Change, Tenure, Value Chains and Emerging Issues.* The Earthscan Forest Library. New York: Routledge, pp. 107–125.

UN Women. 2018. *Rural Women in Pakistan: Status Report 2018.* Islamabad: UN Women.

United Nations Development Programme (UNDP). 2013. *Overview of Linkages between Gender and Climate Change.* New York: UNDP.

World Bank. 2018. *Labour Force Participation Rate for Women.* World Development Indicators. The World Bank Group. https://data.worldbank.org/indicator/SL.TLF.CACT.FE.ZS. Accessed on 25 July 2019.

World Food Programme (WFP). 2018. *Climate Risks and Food Security Analysis: A Special Report for Pakistan.* Islamabad: World Food Programme.

Zaman, Qamar, Mahmood, Arif, Rasul, Ghulam and Afzaal, Muhammad. 2009. *Climate Change Indicators of Pakistan.* Islamabad: Pakistan Meteorological Department.

3 Gendered vulnerabilities in *Diaras*

Struggles with floods in the Gandak river basin in Bihar, India

Pranita Bhushan Udas, Anjal Prakash, and Chanda Gurung Goodrich

Introduction

A *diara,* from the word *diya* (an earthen oil lamp), is an area where a *diya* is never lit. Here it symbolizes villages located inside the embankments of the floodplains of the Gandak river in Bihar. In a wider sense, *diara* is used to indicate people living in abject poverty and facing multiple vulnerabilities due to frequent flooding of a river, here the Gandak. It is a meandering river that may change its course unpredictably (Choudhary et al. 2019). The Gandak enters India from Nepal where it is known as Narayani or Gandaki. It flows south through seven districts of Bihar and two districts of Uttar Pradesh before joining the Ganga at Hazipur in Bihar. More than 34 million people live in these nine districts (according to the 2011 Census), most of them in flood-prone areas.

People living in diaras in the context of a changing climate may experience either too much or too little water, so it is important to understand their vulnerabilities. The two major climatic parameters – floods in the rainy season and relative droughts in the summer – affect the livelihoods and well-being of the diara inhabitants. The impact of these stressors as well as the capacity to cope with them varies according to caste, class and gender, due to the dissimilar access to resources across social groups. Understanding people's struggles to survive in difficult circumstances is crucial for an understanding of their adaptation and survival strategies in response to various stressors (Gilson 2013). This chapter, which analyzes gender-based vulnerabilities of people living in diaras with respect to their social, economic, environmental and political resources, seeks to inform government policies and plans for disaster management and the promotion of adaptation strategies, as well as to contribute knowledge to ongoing development efforts and to academic discourses on vulnerability.

This chapter takes the view that understanding the vulnerabilities of people from their own perspectives provides a realistic picture of the realities on the ground. It uses participatory assessment tools to understand climatic and socio-economic drivers and conditions contributing to the vulnerabilities of diara residents. Data were collected using various methods, including focus group discussions and interviews, participant observation and transect walks. Both physical aspects of the river, rainfall and the embankment, and socio-economic aspects including social position and gender, were considered in the analysis. The fieldwork was conducted between February and July 2016 in four different villages in the West Champaran district of Bihar. The four village contexts were, respectively, recurrent floods, new flood zones, floods triggered by infrastructure development and remoteness from government agencies. All the villages were located within the Gandak river embankment.

This chapter is divided into five sections. The first section describes the floods in the Gandak river basin, and the climatic and socio-economic issues faced by the villagers. The second section outlines the conceptual framework for gender-based social vulnerabilities from the existing literature. Empirical evidence from the field is drawn on in the third section to show how gender-based vulnerabilities are an outcome of gender-based norms and values with respect to the use, access and impact of environmental, economic, political and social resources. The next section focuses on the coping and adaptation strategies of the people. The chapter concludes by highlighting our contribution to the literature, especially in terms of research methodologies and conceptualizing the creation of gendered vulnerabilities as well as building resilience and adaptation to climate change.

Diaras: A land knee deep under water

Bihar is prone to flooding: 73 per cent of its land area and 76 per cent of its population are perpetually under this threat (Kumar and Sahdeo 2013). Out of 38 districts, 28 are at risk (Figure 3.1). Mountain-fed river systems like the Kosi and the Gandak, a foothills-fed river system like the Bagmati, and plains-fed river systems like the Burhi Gandak cause floods in the state every year (Sinha and Jain 1998).

Such floods are not always natural but are often caused by a breach in the embankments built to protect people from inundation or by a diversion made in connection with infrastructure development such as roads. In government records the floods of 2004, 2007, 2011 and 2013 are termed natural disasters, whereas the 2008 flood was regarded as breach induced (Government of Bihar 2016). The most recent flood, in 2017, was devastating, breaking a nine-year record of deaths caused by floods. In West Champaran district alone, 42 people died and around 46,000 were affected (Government of Bihar 2017).

Floods are a hurdle to growth in Bihar. The 2011 *India Human Development Report* ranks Bihar in the bottom five states according to the Human Development Index (Government of India 2011). According to the report, the poverty headcount in Bihar in 2009–10 was about 54 per cent in comparison to about 30 per cent for the whole of India (UNDP 2016). West Champaran is one of the poorest districts of Bihar with 77 per cent people living in poverty in 2004–05 (Chaudhuri and Gupta 2009). In rural areas, 81 per cent of households lived below the poverty line (Pankaj and Mishra 2008).

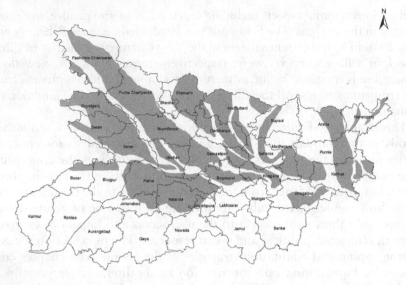

Figure 3.1 Flood zones in Bihar.

Building Material and Technology Promotion Council, Government of India and UNDP (2016), adapted by the authors from http://www.disastermgmt.bih.nic.in/Map/images/FloodZoneBig.gif.

The floods caused by the Gandak are one of the reasons for the under-development of West Champaran. The Gandak drains into India through West Champaran, carrying a heavy load of silt and sediments, which causes the river to change its course every year (Sinha and Jain 1998). Of the 1,491 villages in 18 blocks of West Champaran, 156 villages in 17 blocks are regularly affected by floods (DAWC 2013). According to the 2011 Census, 118 villages in the district are uninhabited and many of them have been abandoned, primarily due to the river's frequent changes of course.

The floods are caused not only by localized rain but also by rainfall in the upstream areas of the basin (Ghosh and Mukhopadhyay 2014:3). When-ever there is excessive rain, the gates of the Balmiki barrage on the Gandak are opened to allow the water to flow, to prevent damage in upstream areas and to the barrage structure. The sudden increase in the volume of water in this stretch of the river causes flooding. Failure to communicate infor-mation about the opening of the barrage well in advance leads to loss of life and property. Floods caused by the changing course of the river due to deposition of sediment can also inundate habitats.

Flooding in the Gandak generally causes more damage to property than loss of life, since the floods in this area are of the slow onset riverine type, allowing people to anticipate rising water levels and evacuate in time, except in areas of flash flooding. The major challenge here is that the floodwaters inundate the land for at least four months during the mon-soon. Loss of property and mobile lifestyles are challenging for people, some of whom take shelter on public land such as embankments, in highland areas, or on rented land.

Women, children and the elderly are highly vulnerable to hardship during floods due to socio-cultural norms such as *lajja* (shame), forced mobility, the special needs of women in certain situations and the increased workload of managing water and other local resources to look after the family (Mehta 2007). The lack of local economic opportunities forces men to migrate to work outside the village, while children, women and the elderly are left behind in the flood-affected areas throughout the year (DAWC 2013).

Troubles continue after the monsoon, when the flood waters drain, with severe water shortages and drought in the winter. Bihar experienced significant droughts in 2006, 2009 and 2010 (Government of Bihar 2016). Despite these hardships, people live in the diaras because they have no alternative. For some, the land they used to live on became diaras because the river changed its course. For others, their land fell inside embankments, for example, the Pipara-Piprasi embankment (Figure 3.2). For many others, diaras are the only place left for them to live, because it is impossible to claim land rights elsewhere, given the feudal social structure of Bihar where access to land is skewed.

The socio-economic context and changing climatic conditions in West Champaran

'Crime, caste and cost [bribery] are push factors for poverty and vulnerabilities in Bihar' is a local aphorism. Together with climatic stressors, vulnerabilities for some groups of people in West Champaran are rooted in histories of social discrimination and domination.

West Champaran shares a border with Nepal and criminal activities in the area are facilitated by an easy escape across the border. Land ownership is skewed: much of the land is in the hands of a few landlords owning more than 20 ha, who align across a strict caste-based hierarchy. According to the Agriculture Census of 2015–16 (Figure 3.3), 97 per cent of landholders are small and marginal, owning 74 per cent of the land in West Champaran, while the remaining 3 per cent own 26 per cent, i.e., almost one-third of the land (Government of India 2012). Historically, strong caste-based disparities, the *zamindari* system and colonization by the British of indigo plantations have created a structure of domination and inequality in the area. Although the *zamindari* system was abolished and the Land Ceiling Act came into force after independence, skewed land relations persist, indicating the continuation of socio-economic inequality. There is a lower proportion of women landowners in Bihar than the national average of 11 per cent (Rao 2011). A recent study in two districts of Bihar found that only 7 per cent of women are landowners (Golder 2017).

The 2015–16 National Health Survey of West Champaran district shows the low status of women in rural areas (Government of India 2017). The literacy rate for women was 40 per cent, far less than for men (70 per cent).

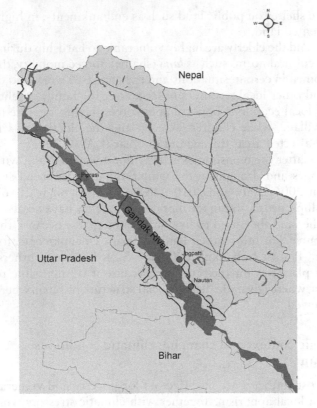

Figure 3.2 Pipra-Piprasi Embankment in Gandak River. Study area plotted with estimation, with respect to GPS map in HI–AWARE (2017).

Source: Adapted by authors from Government of Bihar (2015), HI–AWARE (2017).

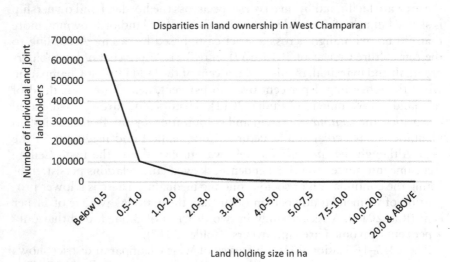

Figure 3.3 Disparities in Landownership in West Champaran.

Agriculture Census, 2016.

Only 12 per cent of women had ten or more years of schooling. Early marriage was widespread. In the rural areas 23 per cent of girls aged 15–19 were either pregnant or already married with children. In addition, 42 per cent of children under five were underweight. Anaemia was widespread: 63 per cent of children aged between 6 months and 5 years, 57 per cent of women and 27 per cent of men between 15 and 49 were anaemic. The caste system continues to thrive in society, regulating all social practices related to birth, marriage, death and various other rituals. Caste-based discrimination still prevails. A strong son preference is common. In rural areas, the sex ratio is 936 women per 1,000 men while the child sex ratio (children under five) is 870 girls per 1,000 boys according to the 2011 Census (Government of India 2017).

Against this backdrop of continuing structural inequality and deprivation, changing climatic conditions are starting to impact West Champaran. An analysis of monthly rainfall data by the Indian Meteorological Department (IMD) during 2013–15 indicates a shortfall in rainfall in West Champaran. Climate data for the area downstream of the Gandaki basin, including West Champaran, over 1981–2010 shows a slightly declining trend with respect to the intensity of rainfall for different thresholds, but none of them were significant (HI-AWARE 2017).[1]

The climatic modelling data found a significant increase in temperature over the last 30 years. A scenario for 2050 reveals that regions downstream of the basin, including West Champaran, will be warmer. Currently, the minimum temperature in the Gandaki basin tends to be below 30° Celsius (C). Under the highest Representative Concentration Pathways (RCP) scenario calculated by the Intergovernmental Panel on Climate Change (IPCC) – the 8.5 RCP scenario – areas around Gandak, including West Champaran, will experience heat stress for 1–20 days a year with a minimum temperature of 30° C and possibly more drought (HI-AWARE 2017).

Understanding gendered vulnerabilities

The IPCC defines vulnerability as:

> the extent to which a natural or social system is susceptible to sustaining damage from climate change. Vulnerability is a function of the sensitivity of a system to changes in climate (the degree to which a system will respond to a given change in climate, including both beneficial and harmful effects) and the ability to adapt the system to changes in climate (the degree to which adjustments in practices, processes or structures can moderate or offset the potential for damage or take advantage of opportunities created, due to a given change in climate). Under this framework, a highly vulnerable system would be one that is highly sensitive to modest changes in climate, where the sensitivity includes the potential for substantial harmful effects, and one for which the ability to adapt is severely constrained.
>
> (Watson et al. 1998:1)

Adger defines social vulnerability as "the exposure of groups of people or individuals to stress as a result of the impact of climate change" (1996:5). Providing evidence of vulnerabilities due to famine, natural and climate-related hazards, Adger emphasizes the underlying causes of social vulnerability at individual and collective levels. Ciurean et al. (2013) locate social vulnerability as a bottom-up approach in contrast to physical vulnerability approaches, which are top down and focus more on future scenarios (Figure 3.4). The authors highlight the effectiveness of adaptation policies that consider the assessment of social vulnerabilities through a bottom-up approach in relation to physical vulnerabilities, and vice versa.

As a subset of social vulnerability, gender-based vulnerability is part of a process that creates differential vulnerabilities for people belonging to different gender-based social categories (Sugden et al. 2014; Goodrich et al. 2017). It is important to understand that gender is not just an indication for women and men; rather, it is considered as part of heterogeneity in gender categories and inter-sectionalities (Ravera et al. 2016).

Gendered life is defined as an organizing principle of social life, creating and ordering relations between people in a hierarchical manner as well as

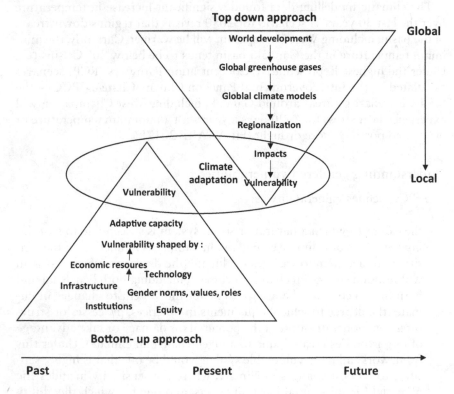

Figure 3.4 Top down versus bottom up approach of vulnerability.

Adapted from Ciurean et al. 2013.

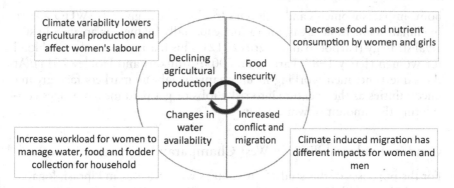

Figure 3.5 Differential gender impact of climate change in an agrarian society.
Complied by the authors.

giving meaning and legitimization with respect to performance, resource allocation and social practices to certain sex-based groups belonging to specific social categories (cf. Harding 1986). The way different social groups are organized shapes the gender-differentiated possibilities of responding to climatic stressors.

Denton (2002) explains that gender inequalities continue to exist in terms of access to land, control over resources, ability to command and access to paid labour, capacity and strategies for income diversification, as well as time spent on agricultural or forestry-based activities. Gender dimensions of vulnerability, in the context of a changing climate, derive from differential access to social, economic, environmental, political and technological resources and exclusion from decision-making processes that help in coping and/or adapting to these changes (Vincent 2004:41). Access to various livelihood resources is shaped by gender-based social relations reflected in gender-based norms, values, roles and responsibilities. Greater mobility, raising voices and decision making within the family and in public become possible for certain gender-based social groups in response to disasters and stressors, due to a different entitlement to resources and because of social relations that influence access to and control over them (Kabeer 1994; Sen 1999).

The literature on the gender-based impacts of climate change in an agrarian society highlights the impact of water availability and associated resources. It also underscores the gender-based dynamics with respect to the division of roles, increased workload and resulting vulnerabilities (Dankelman 2010). Figure 3.5 summarizes the differential gender impacts of climate change in an agrarian society. What we know less about is the gendered vulnerability and adaptation of people in agrarian contexts like the diaras, where land as basic source of livelihood is destroyed or lost with recurrent floods (Choudhary et al. 2019).

Flood and drought have an adverse gendered impact on health (WHO 2014). In the South Asian context, reduced food productivity and storage at home impacts women's and children's food intake negatively (WHO 2014). In climate-vulnerable areas, migration, especially by men, is an age-old survival strategy (Farbokto and Lazrus 2012). This has increased the workload for women (Jetley 1987; Paris et al. 2005; Adhikari and Hobley 2011). At the same time, men working in unorganized labour markets face greater uncertainties as they venture into risky jobs or get paid meagre wages considering the amount of work put in.

Gendered vulnerabilities in West Champaran

For the purposes of this study, four villages were selected in Piprasi, Jogapatti and Nautan Blocks in West Champaran District. These villages illustrate the differential nature of, and reasons for, gender-based vulnerabilities of people living along the Gandak (Figure 3.2). Environmental resources are fundamental for agrarian livelihoods, including for the diara inhabitants. The whole area is inundated with water during the monsoon. The waters dry up in winter, resulting in water scarcity. This has implications for the use of land, the basic resource for farming. In our study area, Piprasi Block is located farthest from the district headquarters, Bettiah. The latter, and most of West Champaran, is located on the right bank of the river, whereas Piprasi is located separately on the left bank. Some hamlets in this block become marooned on islands, inundated for three or four months during the monsoon and post-monsoon seasons. They can only be reached by boat.

Another study village in Jogapatti block, which consisted of about 550 households, used to be safe from floods. However, a shift in the river course in 1999 flooded the village; in 2000, the river flowed through the village. In 2001 inundation had become so severe that people began to move to higher places nearby during the monsoons. From 2002, those who could afford to purchase land in safer places started to maintain two homes, one in the village and the other at a higher location to which they could move during floods. The areas inundated by the river expanded every monsoon, turning the village into a diara.

The third village studied, in Nautan block, is a settlement on an old embankment that is government land. A huge flood in 1980 displaced people from their native village. After moving to ten different places in the last 20 years, the villagers settled on this old embankment but had no proper land entitlement. During our field study in 2016, the settlers told us that the floods of 2009 and 2013 had been the most difficult. They also thought that floods had occurred less frequently in the last 10 years. However, the severe floods of 2017 inundated the land for months again, forcing villagers to take shelter on a new embankment.

The fourth study village, further downstream in Nautan, faced several floods between 1974 and 2010. Increased flooding caused famine here. From 2010 onwards, following construction of a road that in practice acted as an embankment, the flood watershed stopped coming inside the village. However, in view of pressure on the road from the river, construction of a new flush gate is planned, which is likely to lead to the village flooding again.

One significant consequence of flooding in all the villages was that farming was challenged by land degradation. Yet, those who had lost most of their land did not lease land from others. One reason given was that returns from farming were no longer assured, as the uncertain rainfall and dryness in winter lead to crop failure. Irrigation using a (rented) groundwater pump is imperative in the drought season, but the poorest people in the diaras cannot afford to do so. In Nautan, of the 106 households living on the old embankment, only two families own land and only 17 had opted for sharecropping in 2016, while the remainder lived on daily wages. Rice is grown in the rainy season, and wheat and maize in winter. A man from a Dalit caste explained, "If paddy planted in the field gets inundated for more than a week or two, the crop completely fails. Similarly, if we fail to irrigate during the dry period, the harvest will be minimal."

As a coping strategy, both men and women opt for daily-wage work. Most of the women work in the village, enabling them to cope with their family responsibilities, such as cooking, child care and other household tasks. Men can work outside the village. "Since we have no other way of earning, wage labour provides a more secure income to meet our daily needs compared to the risks associated with farming", explained a woman from a *Mahadalit* caste.

"We need more men in our homes"

People living in the diaras belong to the most marginalized social groups. For instance, the hamlet on the old embankment in Nautan block consists of 49 per cent *Mahadalit* (socio-economically marginalized *Dalit* caste categorized as *Mahadalit*) belonging to *Mushar* and *Chammar* castes, 38 per cent other backward castes (OBC) belonging to *Hazam, Barai, Bind, Dhobi* and *Malah* castes, 11 per cent *Yadav* (backward caste). The other villages have similar caste compositions. Skill-based, non-farming castes, such as the *dhobis* (who traditionally do laundry work for a living), often possess no land at all.

The society is deeply patriarchal and regressive social practices, such as dowry giving, are prevalent in all the villages and among all castes. Having a daughter is considered a burden rather than something to celebrate. This is because of the social perception that more men means more hands to go out of the village to earn. Girls are considered a liability for whom a dowry has to be provided. Similarly, men in the diaras have difficulty finding a bride from outside the diara. They are a less preferred choice for villagers outside. Marriage is more or less compulsory and occurs when the couple are in their teens. Marriage carries expectations that producing more children, especially sons, can bring in income for the family. It is common for people here to have as many as five children, in comparison to 3.27 as per the 2011 Census data (Govt of India 2011).

In order to understand local perceptions of vulnerability, we asked people living in diaras "Which household is the most vulnerable during a flood?" The answer revealed the societal gender biases: "Households with more daughters are the most vulnerable." "And why are they the most vulnerable?" we asked. The response was that such households have limited social and financial capital to respond to floods. Social and financial capital are crucial for coping with stressors (Adger et al. 2003).

The search for the most vulnerable households led us to one house where a widow lived with her five daughters, and a second where the woman's husband had migrated to Punjab for work. This couple had a daughter with a disability. The husband would visit his wife and daughter once every two or three years and provide limited support. In the third house a widow lived with her son who was ill. All these women managed their family's day-to-day food requirements through waged labour within the village. The problem was that the waged labour was not available all year round, especially during the four months when the agricultural land was under water. Most of these families borrow from moneylenders to survive the four months of flooding, but with repayments spread across the year they find themselves in a perpetual cycle of borrowing and repaying.

How do people protect themselves during times with major floods? With riverine floods, houses are under water for a long time, leading to potential structural collapse. The houses are made of bamboo huts with mud and cowdung plaster and are fragile when they remain in water for long time. Hence, these huts are protected by a tripod-like structure (locally known as *machhan*) attached to nearby trees. At the top there is a seating area, and here the men sit, supporting the structure so it does not collapse. Preparing a *machhan* and guarding household goods are considered men's work. While this *machhan* can easily be used by men as a toilet, it does not provide the necessary privacy for women. So, women trying to guard their hut and goods face more difficulties. Also, gender-based norms discourage women from performing these tasks; consequently, they tend instead to move to higher places.

As a result, a household with more women members has to rely on men from outside or to accept losses. "When there are more men in a family shifting is easier", remarked a woman in Nautan. It is a challenge to find men workers when the flood waters are dangerously close. Everyone wants to help their own family first. Female-headed households are therefore most vulnerable during this time as they have to rely on other men in the community. The dowry practice is another reason for women and households with more daughters to be more vulnerable in the diaras. Using baseline data[2] from 106 households living on the old embankment in Nautan, our third village, a linear trend analysis of the number of loans taken out by a household against the male/female ratio of its family members was undertaken. It showed a positive association between the number of daughters in a family and its loan burden. Households with more male members had fewer loans or no loans at all compared to those with more female members (Figure 3.6). In-depth analyses of ten families from the *Mahadalit* and OBC

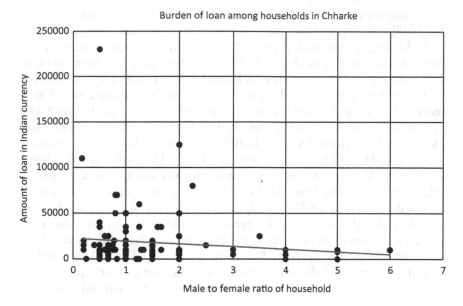

Figure 3.6 Loan burden among households in a village in Nautan.

Source: MPA, 2016.

groups who had taken out loans revealed that the loans were used to pay dowries. Dowry was also mentioned as the main reason for borrowing during the focus group discussions.

For people living in the diaras, more unmarried daughters or women in a household means fewer men able to work outside the village and earn cash incomes for the family,[3] as well as less manpower to protect the family's belongings and save lives. Moreover, having more daughters means extra expenses for marriage. If the daughters remain unmarried, then the family has a social stigma to bear. These social factors have made households with more daughters more vulnerable to climatic stressors in the diaras. Our comparative analyses of households led to the conclusion that households with a higher number of women are trapped in multiple vulnerabilities.

Multiple vulnerabilities of work and migration

We analyzed the annual livelihood strategies of the community members in response to climate change. It was found that the male members of all the households in the diaras, from the age of 11 onwards, tend to move out of the village to earn an income for their families. Their mobility peaked from June to November (DAWC 2013). More recently a few young girls have tended to join them in going outside the village to earn. However, families consider this an enforced trend brought about by food scarcity rather than a welcome livelihood strategy.

The women who stayed back observed it was only a temporary solution that family members, mostly men, were leaving the village for work: "Our men who migrate for work are mostly employed in the unorganized sector like construction and mechanized farming, where they are often involved in hazardous work. Besides, there is limited health cover in case of illness."

There were reports of the disappearance of family members, and of accidents and disabilities – and for some even death – in the case of those who were forced to return. A family identified as most vulnerable in Nautan had a son who had returned with a disability acquired while he was working in Amritsar. He had been bedridden for the last 15 years. Care was being provided by his wife and mother, adding to their workload. A woman in Bairi, where villagers from the Jogapatti block had taken shelter during the floods in 2016, said, "The mother of an 11-year old boy who went to work in Nainital is still in shock after her son died while working." For the women who remain behind, male members leaving for work in uncertain circumstances causes worry and even mental illness. The lack of employment opportunities in the village itself and going out to work in the informal sector has increased the vulnerability of men. A 60-year-old man said:

> We own two *katha*s (1 katha = 0.06 acres) of land. Eighteen of our family members are dependent on this land. Our sources of living are sharecropping, four goats and a cow on shared ownership and earnings from bamboo products. Most of our cash income comes from my son who is a casual labourer in Haryana. He left the village 12 years ago. Every month he used to send half his earnings (of about 3,000–5,000 rupees) which we spent on food, healthcare and repaying the debts (Rs.75,000) incurred for the marriage of four daughters. To add to this stress, while working in Haryana in 2014, my son broke his leg in an accident. It took him a year to recover for which we took out a loan of Rs. 100,000 for his medical expenses. He recovered and left for Patna for work.

For people living in diaras, employment opportunities are limited. The decline in farm productivity due to climatic stressors has forced them into waged labour in the village and outside where pay is higher. Members of the same family may find themselves living in two different places. Men who work in the unorganized sector are vulnerable due to lack of health insurance. Women who remain behind are vulnerable due to gender biases. Household work and caring responsibilities constrain women and restrict their involvement in work outside the village. Women doing waged labour in the village can work a bare 10 days a month, earning Rs. 50 a day for 10 hours of work, because the labour market in the village is limited. Food security is a major challenge, the effect of which is reflected in the overall health status of rural women and children in West Champaran. It is worst in the diaras, where our study team observed malnourished children and helpless families with disabilities.

People in the diaras living on public land suffer from an additional vulnerability: not owning land disincentivizes them from investing in improving their homes. A government-employed *Mahadalit* community worker observed, "If we knew we were allowed to stay on this land forever, we could start investing in improving the structure." Showing us her bamboo hut with a thatched roof, she remarked ruefully: "This structure hardly withstands a minor flood."

Role of institutions in times of flood

Gender vulnerabilities are also manifest in the availability of and access to government resources during floods. These include health services, drinking water and sanitation, food security and infrastructure development. They are interlinked and together they define social and gendered vulnerabilities in the diaras.

During floods, shallow tube wells or dug wells are widely used for drinking water. However, because these wells source water from a shallow table, they become contaminated. Sanitation is the major issue because of open defecation. Water-borne diseases such as diarrhoea and dysentery are common during the floods and particularly affect children, women with special needs and the elderly (Prakash et al. 2015). Government health centres provide free medication to villagers. However, access to such facilities varies. A village that first became an island and was later inundated now has to depend on boatmen to cross the river to reach these centres. This can sometimes take a whole day, for ferry services may not be available immediately or all the time. In one case, a man came to the health centre in Piprasai by boat in the morning, leaving his sick children at home with his wife. He waited the whole day to be taken back. When the boatman appeared late in the afternoon, he asked urgently: "Are you going back to the village now?" The boatman replied, "No, I am waiting for more people to join. I can make only one trip to the village."

At least it was possible for this man to travel to the government health centre, although it did affect his work. For women in a similar situation, mobility is restricted by their roles as homemakers and carers and by social norms. Hence, women-headed households and families with a majority of women have to depend on their men to access government services, or often never access them. Floods adversely affect food storage in a household. The entire population of the diaras is below the poverty line and is entitled to a certain volume of rice and wheat under the public distribution system (PDS). This is commonly known as *raashan* (ration) here. However, women-headed households and those with many women reported difficulties in collecting their *raashan*. In one family in a recently flooded village in Nautan, the husband had decided to work outside the village, as all their farmland had been cut off by a branch of the Gandak. His wife had started to work on daily wages while also looking after their five children, one of whom was disabled and four of whom were girls. She depended on other men to access her quota of *raashan*, which she found difficult in the absence of male members of the household:

> Arranging the daily cereal to feed the children is difficult after our farmland had washed away. We are entitled to receive 3 kg of rice and 2 kg of wheat from the government. I cannot go to collect the ration. I depend on my neighbour to bring it for us to save money.

Any infrastructure development should take account of how people will be affected. Two cases of infrastructure development that increased vulnerabilities of people in the diaras were observed during our field study. One concerned road construction, which increased the risk of flooding for one of the villages in the Nautan block. The villagers observed, "Fear of flood due to the sudden opening of the flush gate of the road is much higher than the river flood itself. It could be more disastrous than a riverine flood." The other case involved a groundwater pump for drinking water which lacked a suitable mechanism and where the water quality had not been safety checked. Groundwater in a village in Nautan block was found to be contaminated with arsenic (Bhatia et al. 2014).

Before the closed tube-well system, dug wells were the source of drinking water in the village. Since tube wells have been widely adopted by households in the last 10–20 years, due to the convenience of extracting water in water-scarce conditions, people have begun consuming arsenic-contaminated water, implying multiple grave health hazards (Singh and Ghosh 2014). The director of the Mahavir Cancer Hospital in Patna claimed that there is an increased incidence of cancer in rural Bihar, which could be due to consumption of arsenic-contaminated water. Men dying of cancer increases the vulnerabilities of women family members. In one case, after the death of the husband from cancer, the wife was left alone to look after the family. To aggravate matters, expenses for the treatment of her husband had increased the family's loan burden, for which she would now be liable. Remarrying is taboo in these villages.

Conclusion: Surviving and adapting to change in the diaras

This chapter has investigated the level of vulnerabilities of people living in the diaras in the West Champaran district of Bihar, India. The chapter explored access to various resources and facilities and its intersection with differential gender-based norms and values. We found that in situations of perennial flooding, people have survived by adopting multiple adaptation strategies, which we discuss below.

Early marriage, when the girl is still in her teens, is commonly practised and son preference is widely prevalent. Early marriage adds to the labour force at home when the husband has migrated to earn money. Producing many children used to be a characteristic of agrarian communities in the pre-industrial period all over the world, especially to meet labour requirements for farming (LeVine and LeVine 1985). The difference between typical farming families elsewhere and those living in the diaras of West Champaran is that here the expectation of a new-born child is that it will become not only potential farm labour, but any kind of labour to bring in income in cash or kind, either by working in the village or outside it.

The gender division of roles, with women as caretakers of the family and men as cash earners, is prominent in this area. After the 1980s, when road connectivity expanded, the expectation was that sons would work outside the village in the face of losses caused by flooding. But this has contributed to the increased, though differential, vulnerabilities for both women and men. The women are left behind in vulnerable areas for longer periods, while the men venture out to work in the unorganized labour market, often in initially unknown territories. They are forced to work as labourers as they do not have the skills to get a better, higher-paying job. Though the Government of India has some provision for enhancing skills, more emphasis on skill-oriented training and the creation of work opportunities would improve the livelihood security of those living in the diaras. Better incomes would enhance their resilience. In addition, security of land tenure, housing, safe drinking water and improved sanitation are crucial for resilient livelihoods. For people in the diaras all these facilities are as fluid as the floodwater. Displacement due to frequent floods has forced most vulnerable families to live on public land, like the old embankment in Nautan block, without any entitlement. Lack of land ownership is a barrier to further investment in basic needs such as shelter, water and sanitation.

Giving land security to displaced people can go a long way towards reducing gender vulnerabilities. For instance, in July 2016, people living on the old embankment in Nautan received financial support from the government to build houses under the *Pradhan Mantri Gramin Awas Yojana*. This raised hopes for ownership of the homestead land after more than a decade. People spent money in excess of the government grant to construct permanent flood-resistant brick houses with iron doors. These houses belong to migrant families, who could save money.

Lack of clean drinking water and sanitation is another reason why women, children and people with special needs become more vulnerable during floods; lowered immune systems cause them to become ill quickly (WHO 2014). Fresh water availability is crucial during floods (Gopalakrishnan and Cortesi 2011). Organizations like Megh Pyyne Abhiyan promoted rain water harvesting in Bihar, including at our study sites, where people who were practising it spoke positively about it, particularly the sick, the elderly and women.

However, the adoption of new practices was found to be limited. Cash constraints often prevented families living in houses made of local materials and thatched rooves from buying plastic sheets to harvest rooftop rainwater. In discussions with women's groups, it was concluded that portable rainwater harvesting tools, such as a conical flask that could help collect water in a bucket, would be more useful than static, roof-based water-harvesting techniques.

This is because people are forced to be mobile during floods. Promotion of such tools and emergency relief kits could reduce women's workload, enabling them to manage drinking water during floods.

Women from landless and marginalized families, who remained behind after the men migrated, were found to be actively involved in shared animal care. This growing trend of shared cow and buffalo farming was observed in all our study villages. The women explained, "Looking after these animals provides nutritious milk for children and is also relatively less risky than farming. The fodder can be managed from the land left fallow after floods, and water from the tube well or river." Programmes that support women's ownership of livestock in the diaras, instead of sharecropping, could enhance their resilience in coping with frequent floods.

These findings, which are based on grounded study exploring gender-based vulnerability, could contribute to the formulation of gender-responsive policies and programmes in the West Champaran district. It also holds true not only for the other diaras in Bihar but for all such riverine bars in the Ganga-Brahamaputra Barak basin. Adaptation policies and programmes in response to bottom-up investigations of the gender-based dimensions of vulnerability would be more effective than relying on a top-down approach. If the gender aspects of vulnerabilities and adaptation are not considered, climate-related policies and programmes may end up creating larger gender gaps and negative social impacts (Terry 2009). The state as caretaker of its citizens has a responsibility to respond to the vulnerabilities of the most vulnerable and to safeguard them in a gender-responsive manner.

Notes

1 Threshold in this context refers to consecutive wet days and consecutive dry days at the annual and seasonal levels.
2 The baseline data was collected by HI-AWARE's study partner, Megh Pyne Abhiyan, in 2015.
3 Value one indicates equal ratio of men to women in a household; less than one indicates more women and more than one indicates more men. Source: Baseline data (MPA 2016).

References

Adger, W.N. 1996. *Approaches to vulnerability to climate change*. Centre for Social and Economic Research on the Global Environment (CSERGE). University of East Anglia and University College London, UK. Accessed at http://citeseerx.ist.psu.edu/viewdoc/download?doi=10.1.1.662.5892&rep=rep1&type=pdf. Date of access 23 November 2018.

Adger, W.N., Huq, S., Brown, K., Conway, D. and Hulme, M. 2003. 'Adaptation to climate change in the developing world', *Progress in Development Studies*, 3(3): 179–195.

Adhikari, J. and Hobley M. 2011. *Everyone is leaving- who will sow our fields? The effects of migration from Khotang district to the Gulf and Malaysia*, Swiss Agency for Development and Cooperation Nepal.

Bhatia, S., Balamurugan, G. and Baranwal, A. 2014. 'High arsenic contamination in drinking water hand-pumps in Khap Tola, West Champaran, Bihar, India', *Frontiers in Environmental Science*, 2. Article 49. doi:10.3389/fenvs.2014.00049.

Chaudhuri, S. and Gupta, N. 2009. 'Levels of living and poverty patterns: a district-wise analysis for India', *Economic and Political Weekly*, 44(9): 94–110.

Choudhary, S.K., Kumar, R., Gupta, S.K., Kumar, A. and Vimal, B.K. 2019. 'Development of Tall and Diara land for sustainable agriculture in Central Bihar, India', *Current Journal of Applied Science and Technology*, 35: 1–13.

Ciurean, R.L., Schröter, D. and Glade, T. 2013. 'Conceptual frameworks of vulnerability assessments for natural disasters reduction.' In *Approaches to disaster management – Examining the implications of hazards, emergencies and disasters.* IntechOpen. doi:10.5772/55538.

Dankelman, I. 2010. *Gender and climate change: An introduction.* London: Routeledge.

DAWC. 2013. *Hazard, vulnerability and capacity assessment.* West Champaran, Bihar, India: District Administration of West Champaran (DAWC)

Denton, F. 2002. 'Climate change vulnerability, impacts, and adaptation: why does gender matter?', *Gender and Development*, 10(2): 10–20.

Farbokto, C. and Lazrus, H. 2012. 'The first climate refugees? Contesting global narratives of climate change in Tuvalu', *Global Environmental Change*, 22(2): 382–390.

FMISC. 2011. *Flood Report 2011.* Patna: Flood Management Improvement Support Centre, Water Resources Department, Government of Bihar.

Ghosh, T. and Mukhopadhyay, A. 2014. *Natural Hazard Zonation of Bihar (India) using geoinformatics: A schematic approach.* London: Springer.

Gilson, E. 2013. *The ethics of vulnerability: A feminist analysis of social life and ˜practice.* New York: Routledge.

Golder, S. 2017. *Development in Bihar – some policy prescriptions from a gender lens.* Oxfam Policy Brief No 24. Oxfam India, New Delhi. Available online at https://www.oxfamindia.org/sites/default/files/2018-09/Development%20in%20 Bihar%20Some%20Policy%20Prescriptions%20from%20Gender%20Lens.pdf. Date of access 19 November 2018.

Goodrich, C.G., Mehta, M. and Bist, S. 2017. '*Status of gender, vulnerabilities and adaptation to climate change in Hindu Kush Himalaya: impacts and implications for livelihoods and sustainable mountain development ICIMOD Working Paper 2017/3.* Kathmandu: International Centre for Integrated Mountain Development (ICIMOD).

Gopalakrishnan, C. and Cortesi, L. 2011. *Rainwater harvesting the start of Megh Pyne Abhiyan's water journey. Megh Pyne Abhiyan – towards self reliance and access to safe drinking water and secure sanitation in North Bihar.* Delhi, Megh Pyne Abhiyan.

Government of Bihar. 2017. *Flood 2017– a report by the Department of Disaster Management.* Patna: Government of Bihar (published 7 September 2017 in Hindi).

Government of Bihar. 2016. *Roadmap for disaster risk reduction 2015–2030.* Revised draft. Patna: Government of Bihar.

Government of India. 2017. *National Population Health Survey 4 2015–16, District Fact Sheet, Pashchim Champaran, Bihar.* Government of India, Ministry of Health and Family Welfare. Mumbai: International Institute for Population Sciences.

Government of India. 2011. *India's Human Development Report Towards Social Inclusion.* Institute of Applied Manpower Research, Planning Commission. New Delhi: Government of India.

Government of India. 2012. *Agriculture Census of India 2010–11*. Department of Agriculture, Cooperative and Farmers' Welfare, Ministry of Agriculture and Farmers Welfare, New Delhi: Government of India.

Harding, S. 1986. *The science question in feminism*. Ithaca and London: Cornell University Press.

HI-AWARE. 2017. *Socio-economic Assessment Report – Gandaki River Basin, Nepal*. Himalayan Water and Resilience Research (HI-AWARE) International Centre for Integrated Mountain Development (ICIMOD), unpublished report.

Jetley, S. 1987. 'Impact of male migration on rural females', *Economic and Political Weekly*, 22(44): 47–53.

Kabeer, N. 1994. *Reversed realities: Gender hierarchies in development thought*. London: Verso.

Kumar, S. and Sahdeo, A. 2013. *Bihar Floods: 2007 (A Field Report)*. National Institute of Disaster Management (NIDM), Ministry of Home Affairs, Government of India, New Delhi.

LeVine, S. and LeVine, R.A. 1985. 'Age, gender, and the demographic transition: The life course in agrarian societies.' In Rossi, A., (ed.). *Gender and the life course*. New York: Routledge, 29–42.

Mehta, M. 2007. *Gender matters: Lessons for disaster risk reduction in South Asia*. Kathmandu: International Centre for Integrated Mountain Development (ICIMOD). http://lib.icimod.org/record/22175/files/attachment_145.pdf.

MPA. 2016. 'Baseline data of Chharke (Nayatola) Nautan, West Champaran, India.' Internal memo. Megh Pyne Abhiyan and Water Action.

Pankaj, A.K. and Mishra, N.K. 2008. *Baseline survey in minority concentration districts of India- Pashchim Champaran*. New Delhi: Institute for Human Development. Unpublished.

Paris, T., Singh, A., Luis, J. and Hossain, M. (2005, June 18). 'Labour outmigration, livelihood of rice farming households and women left behind: a case study in Eastern Uttar Pradesh', *Economic and Political Weekly*, 2522–2529.

Prakash, A., Cronin, A. and Mehta, P. 2015. 'Introduction: Achieving the desired gender outcome in water and sanitation'. In Cronin, A.A., P.K. Mehta and A. Prakash (eds), *Gender issues in water and sanitation programmes: Lessons from India*. New Delhi: Sage Publishing India, pp. 1–21.

Rao, N. 2011. 'Women's access to land: An Asian perspective.' Expert paper prepared for the *UN Group Meeting 'Enabling Rural Women's Economic Empowerment: Institutions, Opportunities and Participation'*. Accra, Ghana, September, pp. 20–23.

Ravera, F., Martín-López, B., Pascual, U. and Drucker, A. 2016. 'The diversity of gendered adaptation strategies to climate change of Indian farmers: a feminist intersectional approach', *Ambio*, 45(3):335–351.

Sen, A.K. 1999. *Poverty and famines: An essay on entitlement and deprivation*. Delhi: Oxford University Press.

Singh, S. and Ghosh, A. 2014. 'Groundwater arsenic contamination and associated health risks in Bihar, India', *International Journal of Environmental Research*, 8(1): 49–60.

Sinha, R. and Jain, V. 1998. 'Flood hazards of north Bihar rivers, Indo-Gangetic Plains.' In Kale, V.S. (ed.), *Flood Studies in India*. Geological Society of India Memoir, 41, pp. 27–52.

Sugden, F., de Silva, S., Clement, F., Maskey-Amatya, N., Ramesh, V., Philip, A. and Bharati, L. 2014. 'A framework to understand gender and structural vulnerability to climate change in the Ganges River Basin: lessons from Bangladesh, India and Nepal.' IWMI Working Paper 159. Colombo, Sri Lanka: International Water Management Institute (IWMI). doi:10.5337/2014.230.

Terry, G. 2009. 'No climate justice without gender justice: an overview of the issues', *Gender and Development*, 17(1): 5–18.

UNDP. 2016. *Bihar- economic and human development indicators*. New Delhi, India: UNDP.

Vincent, K. 2004. 'Creating an index of social vulnerability to climate change for Africa.' *Tyndall Center for Climate Change Research*. Working Paper 56(41).

Watson, R.T., Zinyowera, M.C., Moss, R.H. and Dokken, D.J. 1998. '*The regional impacts of climate change*' Geneva/UK: Cambridge University Press.

WHO. 2014. *Gender, climate change and health*. Geneva: World Health Organization.

4 Of borewells and bicycles[1]

The gendered nature of water access in Karnataka, South India and its implications for local vulnerability

Chandni Singh

Introduction: water scarcity, social-ecological dynamics and gender

Access to water, whether for agricultural or domestic purposes, is often identified as the most important factor in agricultural productivity and rural wellbeing (Hanjra and Qureshi 2010; Namara et al. 2010). In India, where 68 per cent of the gross cropped area is under rainfed agriculture (Mishra, Ravindra and Hesse 2013), managing fluctuations in water availability and ensuring water access is particularly critical. The government and rural communities are constantly negotiating with and adapting to these water fluctuations. This chapter draws on data from Kolar, a water-scarce district in Karnataka, South India, to demonstrate how changing water availability since the early 2000s (driven by erratic rainfall, groundwater over-extraction, and erosion of traditional water management structures and institutions) has reconfigured household work burdens, and water access and use practices. On a wider scale, privatization of drinking water and decreasing reliance on common water resources, such as streams and tanks, have incentivized competitive borewell digging, which has led to unsustainable water use. These shifts in Kolar's waterscape have implications for household and community capacities to adapt to increasing climate variability and is reconfiguring gendered vulnerability.

Since the 1970s, the Government of India has invested in large-scale water management initiatives such as watershed development (Turton and Farrington 1998; NRAA 2012; Singh 2018) and integrated water management (Saravanan, McDonald and Mollinga 2009). More recently, climate-smart

1 Thank you to Nitya Rao for critical feedback throughout the drafting process and to Amir Bazaz and Divya Solomon for shaping my ideas at different stages. This work was carried out under the Adaptation at Scale in Semi-Arid Regions project (ASSAR). ASSAR is one of four research programmes funded under the Collaborative Adaptation Research Initiative in Africa and Asia (CARIAA), with financial support from the UK Government's Department for International Development (DfID) and the International Development Research Centre (IDRC), Canada. The views expressed in this work are those of the creators and do not necessarily represent those of DfID and IDRC or its Board of Governors.

practices such as solar-powered irrigation (Raitha, Verma and Durga 2014; Bassi 2018) and subsidized drip irrigation (Venot et al. 2014) are being implemented to improve capacity to adapt to climate change as well as secure agrarian livelihoods. Yet these interventions have not responded to the needs of people living with water scarcity, who are constantly negotiating this landscape of scarcity through private, often informal interventions such as digging borewells and hiring private tankers (De 2005; Singh, Basu and Srinivas 2016; Solomon and Rao 2018). Everyday practices of accessing and using water are highly gendered (Truelove 2011; Lahiri-Dutt 2015; Harris et al. 2017), with women typically being primarily responsible for accessing and using water for domestic purposes and men seen as productive water users, in charge of irrigation and managing irrigation infrastructure such as pipes, borewells, etc. (Upadhyay 2005; Harris et al. 2017).

The social-ecological system of which these water access and use practices are a part, has been changing since the early 2000s. Increasing temperature extremes and rainfall variability, changing cropping patterns with a shift to water-intensive cash crops, and the degradation of common water resources are changing the landscape of water availability and hence water access across semi-arid India (Reddy 2005; Amarasinghe, Shah and Mccornick 2008; Kumar 2011; NRAA 2012; Garg and Hassan 2016; Singh et al. 2019). In a race to secure water resources for drinking and farming, these ecological and climatic changes have atomized water resources through competitive borewell digging (Solomon and Rao 2018), shifted gendered work burdens (Rao et al. 2017; Singh 2019) and led to rapid groundwater over-extraction (Rodell, Velicogna and Famiglietti 2009).

Socially, migration and increasing rural–urban flows of people, materials and ideas have loosened community ties and eroded customary access and sharing arrangements, leading to "de-territorialised community spaces and identities" (Robson and Nayak 2010: 275). Increasing encounters with the urban have also shaped aspirations and notions of identity (Singh 2019), eroding the influence of socio-cultural norms and values. This unmooring of the social norms and institutions that encouraged community cohesion has necessitated a renegotiation of the social norms, gendered practices and ideological values attached to place (Rao 2014; Bhagat 2017), with critical implications for how natural resources are valued, accessed, used and shared.

Taken together, these shifts in the social-ecological system have altered water availability and norms of water access and use. While several studies capture how and to what extent water availability is changing or projected to change (Kumar, Singh and Sharma 2005; Sharma et al. 2010; Shrestha et al. 2015), few examine their implications for household vulnerability and water security. There is a rich body of work from feminist studies and political ecology on the gendered practices of water access and use (e.g., Meinzen-Dick and Zwarteveen 1998; Upadhyay 2005; Harris et al. 2017). However, gaps remain in understanding its implications for men's and women's vulnerability to climatic and non-climatic risks, as well as on local climate change adaptation. This chapter attempts to close this gap by using data from Kolar to analyze the implications of changing patterns of water availability, access and use, on gendered vulnerability, at different scales: the household, the settlement and the social-ecological system. This attention to scale follows conceptual shifts in climate change research that call for systems approaches to understanding local adaptation (Adger, Arnell and Tompkins 2005; Cash and Moser 2000; Wilbanks 2015), critical for understanding how wider biophysical shifts and socio-economic dynamics shape gendered individual and household actions.

In this study, vulnerability is understood as the tendency to be adversely affected by climatic and non-climatic risks (Adger 2006). Researchers have shown that vulnerability is socially differentiated, embedded in existing cultural and institutional contexts, everyday norms and rules, and cleaved along existing lines of inequality and marginalization (Cutter 2003; Alston 2013). Empirical evidence highlights that vulnerability is also significantly gendered and intersectional (Carr and Thompson 2014; Rao et al. 2017), where being a man or woman intersects with other factors such as age and ethnicity to shape people's ability to deal with shocks and stressors and attenuate their vulnerability.

This chapter draws on two cases within Kolar: the growing informalization of drinking water access, often carried on bicycles from nearby water sources and towns; and the individualization of irrigation water with an erosion of traditional water management institutions and growing reliance on borewells. Through the imageries of bicycles and borewells, this chapter reflects on what the implications of these shifts in water access and use are for gendered vulnerability of men and women and local adaptation. The rest of the chapter is structured as follows. The next section describes the study site and methodology. The main findings of the study are then discussed, first charting the biophysical and socio-economic changes in Kolar district, then the transitions seen in gendered access and use of irrigation water, and finally the changing nature of drinking water access. The chapter ends with a discussion of the implications of changing water access for gendered vulnerability and local adaptive capacity.

Context and methodology

Context

Karnataka is a largely semi-arid state in South India with 54.8 per cent of its rural population engaged in agriculture as a primary source of livelihood (NSSO 2013). Although the gross state domestic product (GSDP) recorded a growth rate of 8.5 per cent in 2017–18, this was mostly driven by the services sector (10.4 per cent growth) in large urban centres such as Bangalore, with a much lower growth rate of 4.9 per cent in agriculture and allied sectors (Government of Karnataka 2018). Successive Human Development Reports have also highlighted severe socio-economic disparity within Karnataka, especially between northern and southern districts (Government of Karnataka 2015, 2018).

Agriculture is the main livelihood in Karnataka, but it is employing fewer people. Between 1993 and 2005, employment in agriculture fell from 65 per cent to 61 per cent and its contribution to the state's GDP halved from 36 per cent to 18 per cent (Government of Karnataka 2010). While agricultural incomes grew by 4.48 per cent between 2002–03 and 2012–13, non-farm incomes in the same period grew by 5.3 per cent, some of the highest shifts in the country (Ranganathan 2014).

Agriculture in Karnataka faces multiple problems, from pressures such as land degradation and water scarcity, to issues of low and stagnant crop yields, market price fluctuations and high indebtedness. Seventy-seven per cent of all farmers in Karnataka were in debt in 2012–13 compared to the national average of 52 per cent, and indebtedness has increased by 16 per cent since 2002–03 (Government of Karnataka 2014, 2015; Ranganathan 2014). Groundwater in 63 *taluka*s (an administrative division of a district) is overexploited, and critical in 21 *taluka*s (CGWB 2014). The state is chronically drought prone, with 27 of its 30 districts, including Kolar, declared drought hit since 2011. Climate variability is increasing, with records showing higher winter temperatures, more erratic rainfall and monsoonal dry spells becoming more common (BCCI-K 2012; Kattumuri, Ravindranath and Esteves 2015; Kumar et al. 2016; Singh, Basu and Srinivas 2016). Consequently, groundwater extraction for irrigation and domestic purposes is on the rise (Singh, Basu and Srinivas 2016). In Kolar, 42 per cent of the population falls below the poverty line (Census 2011), which is the highest in the state (Government of Karnataka 2014). This provides an opportunity to understand how environmental change interacts with and potentially exacerbates social vulnerability.

Methodology

This study adopts a mixed-methods approach to capture data at intra-household, household and village scales. It focuses on situating household risks, and coping and adaptation strategies, as not only embedded in wider village transitions and social-ecological system dynamics but also shaping intra-household vulnerability and wellbeing. Data were collected through 18 focus group discussions (FGDs), a household survey (n=417), life history interviews (8 individuals in 5 households) and open-ended interviews with farmers (3 female and 6 male) and district government officials (n=7). The fieldwork was conducted during a series of visits between 2015 and 2017 to capture livelihood dynamics and drivers of vulnerability in different seasons.

Within Kolar, nine villages were purposively sampled (see Annex 4 for sampling details) to represent different agricultural patterns and livelihood types identified through extensive scoping visits and key informant interviews. Village profiles were created through transect walks, FGDs, informal interviews and participatory mapping (see Singh, Basu and Srinivas, 2016 for further details). Within the chosen villages, households were randomly sampled to represent different landholding classes, castes, household head-ship and asset poverty categories.

The FGDs were conducted in separate male and female groups. The research team piloted mixed-gender FGDs in an attempt to capture gen-der-based conflicts and agreements, but these pilots proved unsuccessful since in keeping with gendered social norms, women tended to remain silent in front of men. In the gender-differentiated FGDs, we employed three participatory tools – timeline mapping, ranking risks and responses, and stakeholder mapping – to identify actors and institutions people draw on to manage risk. The data from these tools helped develop a longitudinal understanding of social and environmental change in the villages from the 1970s (a significant drought period) to the present day, as well as the net-works people access for livelihoods and risk management.

The household survey was administered to the household head; of the sample of 406 households, 18.5 per cent were female headed. The life history interviews were specifically undertaken to understand intra-household differ-entiation on decision making, work allocation and wellbeing. These were typi-cally undertaken with one male and one female member of the household and followed a semi-structured interview style undertaken over two or three visits.

The research team comprised a lead researcher and four research assis-tants (all except the lead, who was a subject expert, could speak Kannada fluently). The survey was carried out through a local NGO whose local staff (from villages across Kolar) were trained in two 2-day workshops on survey methodology and aims. The survey instrument was in Kannada and all data was checked in real time since the survey was conducted on tablets. The research team accompanied the surveyors on pilot visits to fine-tune data collection. All FGDs, scoping visits and life history interviews were done by the lead researcher with two or three research assistants. Care was taken to have female interviewers for female respondents.

Findings: the changing waterscape of Kolar

Drought, water scarcity and climate variability in Kolar

A conversation in Kolar, famed as the "land of silk and milk" (Government of Karnataka, n.d.), is always peppered with discussions on water scarcity. Topics as wide-ranging as maternal health, agricultural output, political power and property prices are always discussed against the backdrop of water scarcity, increasingly erratic rainfall and plummeting groundwater levels. Kolar is located on the southern *maidan* (plains) of Karnataka and has three minor seasonal rivers (Palar, Papagni and Markandeya) (Fig-ure 4.1). Annual precipitation ranges from 679 to 888mm and more than half the annual precipitation occurs during the south-west monsoon or *kharif* season (Kolar District Office 2015).

Rainfall in Kolar is characterized by uneven distribution and dry spells. Between 1900 and 2005, there was a rainfall deficit in 20–35 per cent of

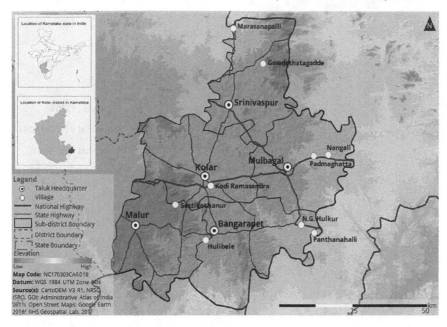

Figure 4.1 Map of study area: Kolar District in South India showing research sites.
Source: IIHS Geospatial Laboratory, 2019.

the years and an excess in approximately 20 per cent of the years (BCCI-K 2012). An analysis of block-wise average annual rainfall (2006–14) shows a steady decrease in rainfall amount over the years (Figure 4.2). Projections estimate a decrease in monsoonal rainfall of 3.57 per cent and an increase of 1.96°C in annual average temperature (BCCI-K 2012). Winter rainfall is projected to increase by 4.9 per cent, which may usher in a change in cropping patterns towards vegetables, as well as a change in varieties (such as sowing short-duration ragi).

From 2011 to 2016, the entire district was declared drought hit (District Collector Office, *pers. comm.*). Meteorological drought (25 per cent less than normal rainfall) was seen across the district; since 2006, none of Kolar's three seasonal rivers or tanks have filled up, and there is no surface run-off, signifying hydrological drought (drying up of surface water).

Changing livelihoods and natural resources in Kolar

Most people in Kolar are employed in agriculture and allied sectors (horticulture, livestock, sericulture) (Figure 4.3) with a larger proportion of women working as agricultural labourers. Within agriculture, 67 per cent of farmers are marginal landholders owning <1 ha land; only 2.8 per cent are medium or large landholders (4 ha and above) (Kolar District Office 2015). In the sampled population from the nine villages studied, 18.6 per

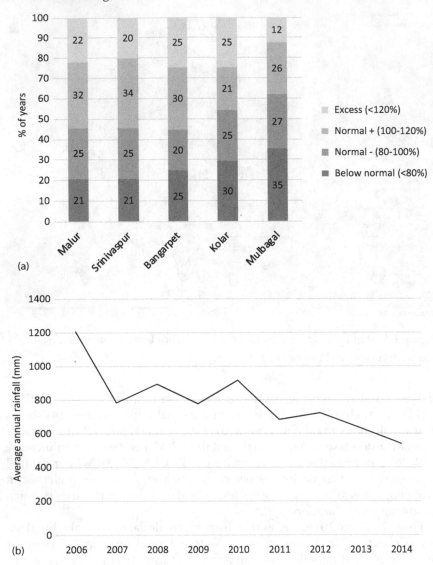

Figure 4.2 Block-wise rainfall patterns in Kolar District, 1900–2005 (top); average
annual rainfall in Kolar, 2006–2014 (bottom).

Source: CRIDA, 2010 and District Office, Kolar, 2015.

cent were landless, 56 per cent were marginal farmers (<1ha land), 15 per
cent were smallholders (1–2 ha land) and only 10.5 per cent were medium/
large landholders (>2 ha). More women (24 per cent) were landless than
men (18 per cent) and the two female-headed households that were medium
landholders belonged to upper castes ('general' category). Landholding and
caste were correlated, with Muslim and Scheduled Castes such as *Bhovis*
and *Adi-Karnataka* having smaller landholdings (average of 1.3 ha) than

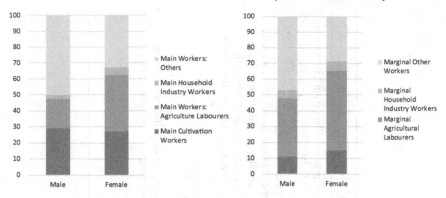

Figure 4.3 Main employment categories in Kolar.
Source: Census of India, 2011.

households belonging to Other Backward Castes (e.g., *Vokkaliga, Kurubas*) and General Categories (average of 2.1 ha).

While the main crops grown are finger millet, groundnut, pigeon pea, cowpea and rice, cropping patterns have changed significantly with a shift away from multi-cropping to mono-cropping of cash crops and horticultural crops such as flowers and vegetables, as revealed by our participatory timelines (Singh et al. 2016) (see Table 4.1). The shift to cash crops has increased farm returns for some farmers, typically those with enough land to grow large quantities and an assured water supply through personal borewells. Across the nine villages we studied in Kolar, these tended to be male-headed households with irrigation infrastructure and farm equipment. This shift to cash crops also has environmental impacts, since they tend to be more sensitive to climate variability and some, such as tomato and sugarcane, require more water. Further, while these crop shifts are lucrative in the short term, they are subject to market fluctuations in Bangalore and Chennai (Technical Assistant, Horticulture Department, Kolar, *pers. comm.*). An illustrative example is of Kolar's key cash crop – tomatoes. In 2017, tomato prices crashed, falling from ₹13/kg[1] in April to as low as ₹2/kg[2] in May, triggering district-wide protests by farmers (Kundapura 2017). With an estimated 9,850 acres under tomato cultivation in Kolar, the district suffered unprecedented losses. As a smallholder farmer in Nangli village noted,

> Prices are very erratic. They can range from ₹50–250 for a sack of tomatoes in a space of 3 days. How are we to make any profit? … In our village, about 100 people work in the tomato yard at N. Voddahalli (32 km away) as agricultural labour. With the price crash, they were also not paid.
> (K15 male, FGD in Nangli village)

Thus, tomato price fluctuations not only had a direct impact on agriculture incomes but also had a cascading effect on agriculture labourers, highlighting how mono-cropping can exacerbate farmer vulnerability.

Table 4.1 Changing cropping patterns and practices in Kolar

	Buchanan (1807), Rice (1897)	Shiva et al. (1981)	Present (KIIs, pers. obs.)
Main crops grown	– Dry seeded rice – Irrigated sugarcane – Finger millet – Legumes (lablab beans, pigeon pea, chickpea) – Red and white sorghum – *Baragu* (fodder)	– For consumption: millets (ragi, sorghum) – Pulses (red gram, broad bean, cowpea) – Oilseeds (horse gram, sesame, niger, groundnut) – Cash crops: Eucalyptus	– For consumption: millets (ragi), rice, maize (very little); pulses (tur, cowpea, broad bean, horse gram); vegetables (potato, tomato, beans, capsicum, cauliflower) – Cash crops: oilseeds (groundnut, sesame, mustard); eucalyptus; sugarcane; fruits (mango, banana, sapodilla, papaya); vegetables (tomato); flowers (rose, marigold, aster)
Agricultural calendar		– Rainfed: finger millet, sorghum	– Rainfed: millets – Irrigated: sugarcane
Cropping type	– Intercropping of millet with legumes; rotations of ragi and legumes; or – Single ragi crop followed by chickpea or sesame.	– Mixed cropping of millets with legumes – Pulses and oilseeds grown with finger millet in rotation	– Mixed cropping of millets with legumes – Mono-cropping of horticultural crops

Source: Author compilation.

Changes in land use and cropping patterns

Most land in Kolar (60 per cent) is under cultivation while 11 per cent is *gomala* land (community-owned land under the *Gram Panchayat* set aside for grazing). Of the cultivated land, 50 per cent is devoted to agriculture, 44 per cent to horticulture and 6 per cent to sericulture (Figure 4.4). Of the net sown area, 17.2 per cent is irrigated while the majority (82.8 per cent) is rainfed (BCCI-K 2012).

However, this categorization masks issues such as the exploitative planting of *nilgiri* (eucalyptus) on personal and *gomala* land, and the shift towards water-intensive and climate-sensitive cash crops such as tomato. The increasing cultivation of *nilgiri* on private and public land over 20–25 years has resulted in loss of soil fertility (Shiva, Sharatchanra and Bandyopadhyay 1981) and increased pressure on groundwater (Joshi and Palanisami 2011). A 2011 study of 21 villages in Kolar district demonstrated that borewells with nearby eucalyptus plantations were deeper due to lower water levels. Further, yields of borewells located within 1 km of eucalyptus plantations fell by 35–42 per cent over 3–5 years (Joshi and Palanisami 2011). *Nilgiri* was introduced as a fast-growing economically useful species under a World Bank social forestry project in the late 1960s and early 1970s, but its negative impacts on local water and soil systems have been reported since the early 1980s (Shiva, Sharatchanra and Bandyopadhyay 1981). In the following decades, 13 per cent of Kolar's agricultural, predominantly ragi-growing land was converted to eucalyptus plantation, with severe implications for local food and nutritional security. At the time of data collection, despite the negative impacts of eucalyptus, several households (especially those where the household head had migrated for work) reported growing it on their personal land since it did not require upkeep and labour.

The implications of eucalyptus for farming and local livelihoods are illustrated through the following quotes:

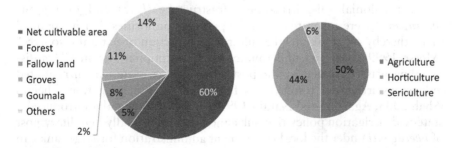

Figure 4.4 Types of land use in Kolar (left) and area under different types of cultivation (right).

Source: District Commissioner Office, Kolar, 2017.

Nilgiri (eucalyptus) has been grown since the 1980s. We grow it even though it is water intensive because we can't do anything else. We plant it and then go for other work to the town.

(K30 village scoping visit, Malur block)

Nilgiri should be banned in the entire district. One *nilgiri* tree requires 6–12 litres of water per day. The tree takes 5–6 years to mature and it is difficult to eradicate once sown. The species ruins the surrounding fields as well, making them unfit for other crops. Every landowner has at least 30–40 trees. The trees are cut and sold to agents for making paper. The farmer gets Rs. 3000–4000 per 1000 kg of wood.

(K28 village scoping visit, Bangarpet block)

Institutional shifts in water management: from the kere system to individual groundwater extraction

Drought and water scarcity are not new to Kolar – early District Gazetteers note that "… the rainfall being scanty and the rivers and streams dry for a large part of the year, the area is, for the most part, devoid of vegetation, and scarcity conditions are very common. There are far more lands under dry cultivation in this district than under wet" (Sri Sathyan 1968: 18). Limited water resources were used ingenuously through the *kere* system – a network of manmade water tanks that exploited the natural slope to form a system of cascading rainwater harvesting structures (Shah 2003). Further, water access and use, especially for irrigation, were carefully regulated through the *neeruganti* (village waterman), who was typically a lower-caste man from the village in charge of water distribution and the maintenance of physical water structures (Somashekara Reddy 2007). The *kere* and *neeruganti* systems were embedded in and shaped by local socio-economic contexts, and hence have also changed over time.

Under colonial rule, irrigation infrastructure (*kere*) and institutions (*neeruganti*) were separated from the political authorities that maintained them, thereby overlooking "the cultural construction of natural resources" (Mosse 1999: 304) and undermining these traditional systems. This dismantling was continued by the post-independence state, which neglected indigenous irrigation resources in favour of modern irrigation schemes (Shah 2003; Agarwal and Narain 1997). In 1962, Karnataka introduced a state-wide irrigation policy that subsumed the traditionally hereditary post of *neeruganti* under the local government administration for large tanks. In villages with small tanks, the *neeruganti* system was dismantled and slowly lost its prominence (Somashekara Reddy 2007). While in some areas in Karnataka, *neeruganti* continue to function, their influence is sharply curtailed (Baumgartner and Hogger 2004) and they do not control or participate in decisions about excavating wells or digging borewells. Simultaneously, the centuries-old *kere* infrastructure has also been eroded due to government and private encroachment, the demise of local regulatory institutions to maintain and desilt tanks, and inadequate finances to maintain *kere*s (Agarwal and Narain 1997; Thippaiah 1998; Shah 2003).

Overall, Kolar has seen shifts in farming, from multi-cropping of millet, vegetables and cereals to mono-cropping of climate-sensitive, water-intensive cash crops such as flowers, tomatoes and mulberry. Further, there have been biophysical and climatic shifts in the form of changing green cover, severe groundwater exploitation, more erratic rainfall and higher temperature extremes. Finally, local institutions to manage natural resources, especially water, have evolved over time from communal institutions to more individual access and use practices. To understand the implications of these biophysical, climatic and institutional shifts on water use and access and gendered vulnerability, the chapter now draws on two cases of water access and use for irrigation and drinking purposes.

Borewells and the atomization of water access for irrigation

From a tapestry of artificial rainfed tanks – Sri Sathyan (1968) reports 535 major and 3,300 minor tanks in the (then undivided) Kolar district[2] – very few are still functional today (CGWB 2012; Shah 2003). Instead, Kolar is riddled with borewells and groundwater extraction has reached alarming levels: its underground reservoirs are extracted to 189 per cent (CGWB 2014) and all *taluka*s in the district are categorized as overexploited (CGWB 2012). The percentage of groundwater development rose in most *taluka*s from 2009 to 2011 and all *taluka*s fall into the overexploited category (Kolar District Office 2015). This shift from a culture of water harvesting and storage to one of unmitigated water abstraction has been a slow disaster in the making – the Central Groundwater Board categorizes all blocks in Kolar as overexploited; 33 per cent of borewells dug in 2013–14 have dried up (CGWB 2014).

Compounding this situation of growing water scarcity are state incentives for groundwater exploitation in the form of subsidized borewell construction through the Rural Development and Panchayat Raj schemes. Between 2009 and 2015, 6,087 borewells were dug under this scheme despite the overall success rate (i.e., the chance that a well dug will have water) having decreased from 83 per cent in 2009 to 66 per cent, signalling a reduction in groundwater availability and the inability of rainfall to recharge existing resources. Controlling borewell digging is a constant challenge. As a junior engineer in the district administration explained,

> [The] government has no control over the drilling of borewells. The ideal distance between two wells is 500m but if a farmer has 10 acres of land, he drills a well in one place...if this does not yield water, after some 2–3 months, he digs another well. There are circulars to curb

digging borewells but in the field it is not possible to restrict farmers to drill borewells as he/she sometimes entirely depends on agriculture for livelihood. Authorities such as the Rural Water Supply, Municipal Administration, Geology Centre are all there but if someone is starving, it is best to allow this [borewell digging] instead of letting him/her starve. 80 per cent of the borewells drilled by farmers are not sanctioned/approved by the district authority. Up to about 1000 feet below the ground, water is potable and below this it becomes contaminated so the district administration is providing RO plants. But who is to enforce borewell depths?

(K50 Engineer in Panchayati Raj Engineering
Division, Kolar)

This shift has led to unsustainable water use, driven by the perception that groundwater is invisible and thus limitless, and has changed practices of water access and use (District Horticulture Office, *pers. comm.*). Almost all (95.7 per cent) borewell water is used for irrigation. Across the villages, respondents spoke of the increasingly erratic rainfall which has necessitated the move to groundwater extraction. The following quotes are illustrative:

We have lakes, we have canals, but since 1998–2000, rains have gone down so much that we have no water. If it rains, we grow some vegetables.
(K24 female, FGD in Nangli village)

Borewells were first introduced in the early 2000s and water was available at a depth of 300 feet. They have become very common in the past since 2010. Now (2016) the average depth of the borewells is 1200 feet. Even at this depth, only 1 in 5 borewells strikes water. Open wells stopped yielding water 10 years ago.
(K20 female, FGD in NG Hulkur village)

Farming is a lottery. If it rains, we'll get returns. Now it is difficult to tell when it will rain. We have no perennial source of water. There are 50 borewells in the village of which only ten are functional.
(Male, Scheduled Caste smallholder, Neregatta village)

The quotes above demonstrate how, in areas dependent on rainfed farming, climatic variability has significant implications for farm incomes. With increasing climate variability making rainfall more erratic and insufficient, farmers are shifting to digging borewells. Further, a breakdown of the *kere* system, as discussed earlier, has eroded the critical irrigation safety net *keres* provided in the form of protective irrigation. Thus, farmers have turned to competitive borewell digging, a strategy that is dominated by men, as we shall see, and tends to exclude female-headed households. Unregulated and often invisible, this individualization of water resources has exacerbated Kolar's water scarcity.

In group risk-ranking exercises, both men's and women's groups ranked untimely rains and water scarcity as the biggest risks to agriculture (Figure 4.5). However, risk perceptions were differentiated when it came to ranking risks such as failed borewells (10 per cent of men and no women's groups ranked it as the biggest risk to farming).

As expected, risk perception was mediated by gendered work allocation and normative gender roles. Thus, men reported market issues more frequently than women because men are typically in charge of selling farm produce in towns. Women ranked sowing-related issues such as poor soil quality and lack of seeds which are typically handled by them. Interestingly, electricity shortages were ranked relatively higher by men, arguably

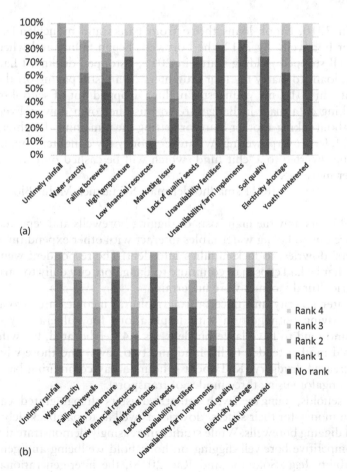

(a)

(b)

Rank 4
Rank 3
Rank 2
Rank 1
No rank

Figure 4.5 Key risks to agriculture ranked by men (top) and women (bottom).

Source: Singh et al. 2016.

because men are mainly in charge of irrigation which requires electricity and a source of diesel to run water pumps in case of power outages.

At an individual scale, successful borewells may also fail after a few extractions: at the time of data collection (September 2016), 33 per cent of borewells successfully dug in 2014–15 and 9 per cent of borewells dug in 2015–16 had stopped functioning (District Collector Office, *pers. comm.*). Such delayed failures also point towards the growing groundwater crisis in Kolar. Socio-economically, failed borewells exacerbate household hardships and inter-generational inequity (Nagaraj and Chandrakanth 1997), since smallholders often take out loans to cover the costs of digging. In Kolar, borewell digging often locked families into cycles of debt, eroding household capacities to invest further in farming or meet family obligations such as wedding expenses, and, as seen in at least seven cases, forced young girls to migrate as domestic workers to raise their own dowries. The following quote elaborates:

> Starting in 2010, I took loans thrice from banks and moneylenders to dig four borewells. By 2012, the borewells began failing and when the last well stopped yielding water in 2013, I stopped drilling. I had also taken loans to marry my four daughters…I am still paying off the instalments. In 2016, my youngest daughter dropped out of school to start working as a maid in Bangalore to raise her own dowry. Even thinking about taking another loan for getting the remaining daughters married off, feels overpowering. Whatever money my daughter is making is being diverted to a 'chit fund' which will be cashed only at the time of her marriage.
>
> (K 41 smallholder male SC farmer, Malur Block)

The quote highlights how the high costs of digging borewells and repeated borewell failure due to falling water tables interact with other expenditures (e.g., daughters' dowries) to lock families into debt. The respondent went on to explain that he had chosen to commute to Bangalore city daily to earn more since agricultural incomes were uncertain.

Growing water scarcity and uncertain agricultural incomes are strong push factors driving outmigration with implications for vulnerability – those left behind tend to have large burdens (as K41 elaborated, his wife and 68-year-old mother tended to his land and two cows) and those who move face increased drudgery (K41 spent 6 hours a day commuting back and forth to Bangalore using three modes of transport).

In some households, young men with unmarried sisters often migrated seasonally to earn money for their sisters' dowries as well as to pay back debts incurred from digging borewells. While studies increasingly demonstrate the impacts of competitive borewell digging on household wellbeing and gendered vulnerability (e.g., Solomon and Rao 2018), the inter-generational lock-ins that borewell-related debts engender remain a relatively unexamined topic. In the life histories conducted in Kolar, three male respondents mentioned repaying loans taken out by their parents to dig borewells and how their income from agriculture was insufficient to repay the loans.

Migration and commuting are becoming significant sources of income in Kolar. Men tend to commute to Bangalore as daily-wage or contractual labourers and migration has been found to be higher in drought years (Kattumuri, Ravindranath and Esteves 2015; Singh et al. 2018). Migration patterns in Kolar are increasingly changing from 'rural-to-rural' to 'rural-to-urban' movement, with more people choosing to work as non-agricultural labourers (Singh and Basu 2019) because of the uncertainty associated with farm labour. Migration patterns were gendered, with 31 per cent of women of the surveyed households migrating compared to 69 per cent of men. Men tended to commute to nearby cities such as Bangalore, whereas women commuted to industrial areas and garment factories in industrial belts near Kolar.

The increased reliance on borewells for irrigation has also changed how men and women farm; and some are excluded from the 'protection' borewell water provides (albeit in the short term). In the villages studied, borewell ownership was always male, which had specific implications for farming in female-headed households. For example, access to borewell water allowed a shift towards crops that were more profitable (e.g., floriculture, tomatoes). While it can be argued that these crops are less climate resilient and hence undermine household adaptive capacity in the long run, in the short run, they allowed borewell-owning households (all male headed) to engage in high-returns agriculture. Access to functional borewells also allowed farmers to experiment more through the perceived protection they provided (in the form of assured irrigation even during erratic rains). Thus, households with functional borewells tended to diversify their crops and exploit new market opportunities more than those without this assured irrigation access. With borewell ownership being markedly gendered, women-headed households were more exposed to monsoon-season dry spells and had less scope to experiment. As a 38-year-old landless, Scheduled Caste female respondent noted, "I am a widow, where will I get the money to dig a borewell? I grow ragi [for which] I can manage with the rainwater. That is all I can do."

On the positive side, Kolar has seen widespread uptake of water-saving technologies such as drip and sprinkler irrigation, which is supported by a 100 per cent subsidy distributed between central and state governments (Chandrakanth et al. 2013). The scheme is implemented through the horticulture department and each beneficiary family can apply for coverage for 5 ha. The scheme has been praised as highly successful, with the highest penetration of micro-irrigation technologies reported in Kolar, where 46.9 per cent of the gross irrigation area and 64.6 per cent of the net irrigated area are covered under micro-irrigation (GGGI 2015). However, the primary data showed that access to micro-irrigation technologies is also heavily gendered: of the total of 417 respondents, only four female-headed households (all belonging to the general category) had access to drip or sprinkler irrigation. In the FGDs, women reported that ownership of drip/

sprinkler irrigation was low among female-headed households because of their small landholdings which did not make the investments financially feasible, difficulties in accessing the subsidies, which required paperwork, and inadequate networks with local horticulture extension staff.

Bicycles and the informalization of drinking water access

Overall, the male- and female-headed households interviewed primarily depend on piped water to their households and public taps for drinking water (Figure 4.6). This water is sourced from borewells on *Panchayat* land and village *kere*. Across the villages, piped water was available for a few hours a day, typically in the mornings and evenings, but the exact timings varied between villages. No female-headed households reported owning a borewell or private well, whereas a few male-headed households did, demonstrating gendered ownership of water infrastructure. Female-headed households reported getting their drinking water from neighbours' borewells (typically those belonging to the same caste), as the village *kere* had over time become polluted and encroached upon, reducing water availability (as discussed further in this section) (Figure 4.7).

In the FGDs, women mentioned that piped water connections are typically provided by the state (though they may be implemented by non-state actors in some cases). However, borewell digging is a household decision, where a household must have the land and financial resources to dig a borewell. In both the FGDs and individual interviews, women across castes discussed how, given the large amounts of money required to dig borewells and having to employ professional diggers, usually from outside the village, no female-headed households got borewells dug for drinking water.

Respondents spoke of the various ways in which they access water for drinking purposes:

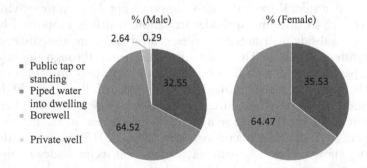

Figure 4.6 Principal source of drinking water in Kolar.
Source: Household survey, 2017–18.

Figure 4.7 Plastic pots for water – public taps exist but are not reliable and involve long waiting times.

Source: Author.

The three *kere*s [artificial lakes] in our village began drying out around 2004. Only two have water now…maybe around 25 tankers (of 10,000 litres each) of water…and it is only the start of summer. We have two borewells for drinking water, which are close to drying out. There is a lot of difficulty during the summer season and we end up spending a lot on buying water from neighbouring towns.[3]

(K10 male, FGD in Hullibele village)

Earlier, water to us was never a cause of concern, but food was. I can't believe I am saying this but now filling our belly is never a problem, but water is. Now everyone in Kolar is organizing and agitating to get access to water. It is our biggest issue.

(K22 female, FGD in Kodi Ramasandra village)

Only three borewells in the village are functioning, but all have hard (non-potable) water because of which we have health problems like pain in the joints. Doctors recommend filtering water, but we don't have filters so we buy water for Rs.15–20 per litre from Malur. But what can the poor people do? We have been doing so since 2011. Sometimes we get tanker water which is supplied by the private borewell owners.

(K15 male, FGD in Nangli village)

The quotes above highlight the dynamic reality of drinking water use and access in Kolar. Biophysical shifts such as lower rainfall, deteriorating common water bodies such as village *kere*s and unequal access to water infrastructure have reduced water availability substantially. For those with the financial resources to dig borewells this has resulted in the individualization of drinking water. For others, it has translated into increased expenditure on buying water. However, purchasing power is uneven, with fewer female-headed household buying water due to the high costs involved and lack of time or transport to travel to towns to buy water.

In the context of growing water scarcity, households have diversified drinking water sources towards private and shared borewells, and state-sponsored and private tankers. In periods of water scarcity (either in drought years or in the lean summer period), accessing water from neighbouring towns such as Bangarpet, Mulbagal and Kolar is becoming a significant reality for many households – men typically buy water in 10-litre cans for Rs.100–150 and use bicycles or motorbikes to ferry them from town to village. In the summer months, it is common to see groups of children carrying an assortment of plastic containers to fill drinking water from nearby settlements with access to *kere*s. Some enterprising men from within the village carry such loads for groups of households for a small fee, creating a livelihood opportunity out of scarcity. Riding bicycles or motorcycles, these 'water carriers' are a critical conduit of drinking water in summer. Conversations with households buying such water highlighted that this water transportation was informal and typically, a young man from the neighbourhood would offer to bring water as part of other errands and charge a small fee such as Rs.10–20 per household. The carriers tended to operate within their immediate social circles of extended family and neighbours, which were always within the same caste. While these informal water carriers are a key way in which households access drinking water in the face of mounting scarcity, this informalization of drinking water access is irregular, costly and governed by dynamics well beyond the village (e.g., water prices in towns or availability of water in neighbouring *kere*s). For example, during a group discussion, a farmer noted:

> About 30 years ago (~1990s) our *kere* covered about 21 acres. Many neighbouring villages took water from it. Increasingly, these lands have been encroached, a lot of sand mining also took place…with people mining sand from the lake bed. Slowly the neighbouring fields also caved in and the miners claimed that as well…now [2015] there is hardly any water for us to drink, let alone other villages.
>
> (K16 male, FGD in Gowdathatagadda village)

In the summer months, when water shortages are acute, the district government provides drinking water through tankers. In 2014–15, the district administration provided drinking water through tankers to 285 villages at a cost of Rs. 33.54 million (District Commissioner Office, *pers. comm.*). Water for these tankers was obtained from government-owned and private borewells across Kolar district. Individuals who sold borewell water received compensation of about Rs.18,000–25,000/tanker (K50 junior engineer, Kolar District, *pers. comm.*). While this water was targeted at poorer households, access to tankers was markedly shaped by people's agency: in the focus group discussions, scheduled-caste women spoke of tanker water being appropriated by influential upper-caste households, with instances of fights and water running out common: "they send tankers but often we are not able to fill our pots...the water falls short and women start fighting... our community has to keep quiet and let the others (upper castes) take their share..." (K28 female, Scheduled Caste respondent from Bangarpet block). This finding resonates with reports of caste-based appropriation of publicly provided drinking water across rural India (Dutta, Behera and Bharti 2015).

In several interviews, women reported relying on male household members to access water. While this did not reduce the time women spent accessing water (since they still queued for government-provided water tanks and at public taps), it has introduced a new form of uncertainty in water access, with women, especially those who were extremely poor, reporting not knowing whether they would have enough water for cooking or drinking the next day. In households with fewer members (e.g., households headed by divorced or widowed women), buying water placed an additional demand on household labour.

For those unable to afford to buy water, drinking unpotable water is the only option, but it comes with severe health implications. As a 36-year-old Scheduled Caste female respondent noted:

> The village *kere* was a source of water for 8 villages earlier[4]...sometime in the early 2000s, rainfall reduced a lot and our *kere* dried up. People dug borewells up to 1300–1500 feet deep...even around the *kere*. Now when we drink the water, we have kidney failure...I've spent almost 12,000 rupees on this. Others have problems in their joints and teeth because of fluoride. I am poor, I cannot buy a water filter.
>
> (K4 scoping visit, Kodi Ramasandra village)

> Most of the woman have knee joint pain and men have kidney failure due to the water here.
>
> (K19 female, FGD Hulibele village)

Water quality is a growing concern in Kolar. High fluoride and nitrate concentration in drinking water (CGWB 2012) has detrimental effects on health with respondents discussing bone deformities and deterioration of teeth. Detailed health and associated expenditure implications were beyond the scope of this research but are reported as critical issues in the region (Chowdhury et al. 2016).

Implications of changing water access for gendered vulnerability and local adaptive capacity

The case of Kolar highlights changing practices of everyday water access and use, charting the shift from reliance on traditional water management institutions (*neerganti*) and infrastructure (*kere*), to the privatization and atomization of water resources (borewells and buying drinking water). On the one hand, these shifts have undermined the local natural resource base (e.g., through groundwater over-extraction, shifts to water-intensive crops and soil degradation) with critical implications for overall sustainability and people's capacity to adapt to changing climatic conditions. On the other hand, informalization and fragmentation of water access has reconfigured water access burdens for men, women and children, exacerbating their vulnerability to erratic rainfall and water scarcity. This chapter argues that these shifts in Kolar's waterscape have not only implications for gendered vulnerability but also wider impacts on local adaptation.

The key contribution of this chapter is that it adds to the growing literature on gender and climate change by demonstrating how changes in the social-ecological system, such as an individualization of water resources or being locked into cycles of debt due to competitive borewell digging, are reconfiguring norms of water access and use, and exacerbating the water insecurity of the most marginalized. While the case demonstrates how water scarcity often creates an opportunity for some to diversify livelihoods (as seen in the emergence of new middlemen transporting water and selling it in villages at a small cost), in the long run, this scarcity can potentially tip systems towards higher social and ecological vulnerability, with differential impacts on men's and women's abilities to cope.

Kolar is symptomatic of wider changes in India's agrarian drylands: (1) natural resource degradation, especially of common property resources such as pasturelands and *kere*s; (2) agrarian transitions towards water-intensive cash crops that are groundwater irrigated; and (3) reactive approaches to seasonal water shortages that typically depend on unsustainable water tankers. Environmental change, as seen in Kolar, is making agrarian livelihoods more precarious, driving cropping changes, out-migration and unsustainable natural resource use. Central to these livelihood strategies is the management of water resources for domestic and irrigation purposes. With the erosion of Kolar's traditional water management system of interlinked *kere*s, the incentivized privatization of water through borewell digging subsidies and the emergence of informal water provisioning through private tankers, water access is increasingly becoming individualized, with increasing dependence on water tanks, individual borewells, private tankers and water bought from nearby towns.

This chapter argues that the implications of these changes resonate across multiple scales (the household, the community/settlement and the larger social-ecological system).

First, at settlement scale, a shift from using common water resources (village wells and tanks) to private water sources (borewells or purchased water) reduces incentives to conserve and protect common resources, hastening their decline and undermining ecosystem health. This erosion of local resources is most acutely felt by women, who often rely on second-order resources such as fodder for small ruminants, fruits from local forest groves, etc. Further, this atomization of water resources has critical equity implications – only those with borewells are able to access protective irrigation, and those with adequate purchasing power are able to buy drinking water from neighbouring towns. Such a shift from communal or individual water access disadvantages the most marginal (e.g., female-headed households or those with fewer male members) and exacerbates existing vulnerability.

Second, at household level, the burden of water collection for domestic purposes, traditionally a woman's role, has shifted to men and young children. Across the nine villages studied, young boys often doubled up as water bearers while returning from school. This has not necessarily reduced the time women spent on water collection since they tended to visit public taps and water tankers to secure any water available. The impacts of such changes on intra-household roles remain to be assessed, but there might be both negative outcomes (higher workloads for young children, health impacts of drinking unclean water) and positive outcomes (potentially reduced drudgery for women). It may also have impacts that are neither negative nor positive but shift intra-household roles such that women are increasingly dependent on male household members for drinking water. For all-women households or those with few or no male members, such structures of dependence may not be possible or might introduce new vulnerabilities.

Third, at a social-ecological system level, private tankers sanctioned by the government and perpetuated by the local water mafia exacerbate the severity of water scarcity in Kolar, necessitating continued reliance on borewells and further groundwater extraction. These spaces are often less available to women, who have to rely on their social capital to access water in these informal ways.

In combination, these behavioural shifts are underpinned by an agenda of extraction rather than augmentation, reducing the capacity and capabilities of people and governments to deal with water scarcity effectively. In the context of projections of increased climate variability, the local government must reimagine its role in ensuring water security and institutionalize creative incentives that discourage over-extraction and reward water practices that do not undermine ecological sustainability and social justice. Further, it is critical to recognize the gendered aspects of water access and use. In the context of increasing climate variability and groundwater over-extraction, understanding the differential impacts on men and women and what it means for their vulnerability is important.

Notes

1 Rs. 13 = USD 0.18 or 1 USD= Rs. 71, as on 15 September 2019.
2 Kolar district was bifurcated into Kolar and Chikkballapur districts in 2007.
3 The respondent is referring to buying water from towns where 5–10-litre cans (typically for Rs.100–150/can) are bought by household members or neighbours using bicycles or motor cycles. These cans are sold by privately owned shops and are typically sourced from private borewells in and around the towns.
4 The *kere* system constituted a network of lakes and tanks that allowed excess water from one *kere* to cascade into others. For details on the system and how it has evolved and been degraded, see Shah (2003).

References

Adger, W.N. 2006. 'Vulnerability', *Global Environmental Change*, 16 (3): 268–281.

Adger, W.N., N.W. Arnell and E.L. Tompkins. 2005. 'Successful Adaptation to Climate Change across Scales', *Global Environmental Change*, 15 (2): 77–86.

Agarwal, A. and S. Narain (eds). 1997. *Dying Wisdom: Rise, Fall and Potential of India's Traditional Water Harvesting Systems*. New Delhi: Centre for Science and Environment.

Alston, M. 2013. 'Women and Adaptation', *Wiley Interdisciplinary Reviews: Climate Change*, 4 (5): 351–358.

Amarasinghe, U.A., T. Shah and P.G. McCornick. 2008. 'Seeking Calm Water: Exploring Policy Options for India's Water Future', *Natural Resources Forum* 32 (4): 305–315.

Bassi, N. 2018. 'Solarizing Groundwater Irrigation in India: A Growing Debate', *International Journal of Water Resources Development*, 34 (1): 132–145.

Baumgartner, R. and R. Hogger (eds). 2004. *In Search of Sustainable Livelihood Systems: Managing Resources and Change*. New Delhi: SAGE Publishing India.

Bangalore Climate Change Initiative – Karnataka (BCCI-K). 2012. *Karnataka State Action Plan on Climate Change*. Bangalore. http://www.karnataka.gov.in/empri/web/EMPRI_KSAPCC_2014-03-27.pdf. Accessed 15 January 2019.

Bhagat, R. 2017. 'Migration, Gender and Right to the City', *Economic & Political Weekly*, LII (32): 35–40.

Buchanan, F. 1807. *A Journey from Madras through the Countries of Mysore, Canara and Malabar*. London: Cambridge University Press.

Carr, E.R. and M.C. Thompson. 2014. 'Gender and Climate Change Adaptation in Agrarian Settings: Current Thinking, New Directions, and Research Frontiers', *Geography Compass*, 8 (3): 182–197.

Cash, D.W. and S.C. Moser. 2000. 'Linking Global and Local Scales: Dynamic Assessment and Management Processes', *Global Environmental Change*, 10: 109–120.

Central Groundwater Board (CGWB). 2012. *Ground Water Information Booklet Kolar District, Karnataka*. http://cgwb.gov.in/District_Profile/karnataka/2012/KOLAR_2012.pdf. Accessed 10 March 2015.

Central Groundwater Board (CGWB). 2014. *Ground Water Year Book 2013–2014 Karnataka*. Bangalore: Central Ground Water Board South Western Region. http://cgwb.gov.in/Regions/GW-year-Books/GWYB-2013-14/Karnataka%20GW%20Yearbook%20%2013-14(1).pdf. Accessed 16 March 2015.

Chandrakanth, M.G., C.N. Priyanka, P. Mamatha and K.K. Patil. 2013. 'Economic Benefits from Micro Irrigation for Dry Land Crops in Karnataka', *Indian Journal of Agricultural Economics*, 68 (3): 326–338.

Chowdhury, C.R., K. Shahnawaz, D. Kumari, A. Chowdhury, R. Bedi, E. Lynch and M. Grootveld. 2016. 'Spatial distribution mapping of drinking water fluoride levels in Karnataka, India: fluoride-related health effects', *Perspectives in Public Health*, 136 (6): 353–360.

Cutter, S. 2003. 'The Vulnerability of Science and the Science of Vulnerability', *Ann Assoc Am Geogr* 93:1–12. doi: 10.2307/973677.

De, A. 2005. '*Mediation in Adaptive Management of Water Resources: Resistance to Borewells at the Grassroots, and Implications for Groundwater Policy Action.*' Working Paper 16. Anand, Gujarat, India: Foundation for Ecological Security.

Dutta, S., S. Behera and A. Bharti. 2015. 'Access to Drinking Water by Scheduled Castes in Rural India: Some Key Issues and Challenges', *Indian Journal of Human Development*, 9 (1): 115–132.

Garg, N.K., and Q. Hassan. 2016 'Alarming Scarcity of Water in India', *Current Science*, 93 (7): 932–941.

Global Green Growth Institute (GGGI). 2015. *Implementation Roadmap for Karnataka Micro Irrigation Policy.* http://www.indiaenvironmentportal.org.in/files/file/Karnataka-MI-Roadmap_Final_WEB.pdf. Accessed 13 May 2019.

Government of Karnataka. 2010. *Karnataka Vision 2020.* Bangalore: Government Press, Vikasa Soudha.

Government of Karnataka. 2014. *Human Development: Performance of Districts, Taluks and Urban Local Bodies in Karnataka, 2014 – A Snapshot.* Bangalore: Planning Programme Monitoring & Statistics Department.

Government of Karnataka. 2015. *Economic Survey of Karnataka.* Bangalore: Planning Programme Monitoring and Statistics Department.

Government of Karnataka. 2018. *Economic Survey of Karnataka.* Bangalore: Planning Programme Monitoring and Statistics Department.

Government of Karnataka. n.d. Kolar. https://kolar.nic.in/en/about-district/.

Hanjra, M.A. and M.E. Qureshi. 2010. 'Global Water Crisis and Future Food Security in an Era of Climate Change', *Food Policy*, 35 (5): 365–377.

Harris, L., D. Kleiber, J. Goldin, A. Darkwah and C. Morinville. 2017. 'Intersections of Gender and Water: Comparative Approaches to Everyday Gendered Negotiations of Water Access in Underserved Areas of Accra, Ghana and Cape Town, South Africa', *Journal of Gender Studies*, 26 (5): 561–582.

Joshi, M. and K. Palanisami. 2011. 'Impact of Eucalyptus Plantations on Ground Water Availability in South Karnataka'. Paper presented at the *21st International Congress on Irrigation and Drainage*, Tehran, 15–23 October.

Kattumuri, R., D. Ravindranath and T. Esteves. 2015. 'Local Adaptation Strategies in Semi-Arid Regions: Study of Two Villages in Karnataka, India', *Climate and Development*, 9 (1): 36–49.

Kolar District Office. 2015. *Kolar District Statistics at a Glance.* Kolar: District Statistics Department.

Kumar, K.S.K. 2011. 'Climate Sensitivity of Indian Agriculture: Do Spatial Effects Matter?', *Cambridge Journal of Regions, Economy and Society*, 4 (2).

Kumar, R., R.D. Singh and K.D. Sharma. 2005. 'Climate and Water Resources of India', *Current Science*, 89 (5).

Kumar, S., A. Raizada, H. Biswas, S. Srinivas and B. Mondal. 2016. 'Application of Indicators for Identifying Climate Change Vulnerable Areas in Semi-Arid Regions of India', *Ecological Indicators*, 70: 507–517.

Kundapura, V. 2017. 'Kolar farmers dump tomatoes on roads as prices crash' in *The Hindu.* https://www.thehindu.com/news/national/karnataka/as-prices-crash-farmers-leave-tomatoes-to-rot/article18348346.ece. Accessed 23 March 2018.

Lahiri-Dutt, K. 2015. 'Counting (Gendered) Water Use at Home: Feminist Approaches in Practice', *ACME*, 14 (3): 652–672.

Meinzen-Dick, R. and M. Zwarteveen. 1998. 'Gendered Participation in Water Management: Issues and Illustrations from Water Users', *Agriculture and Human Values*, 15 (4): 337–345.

Mosse, D. 1999. 'Colonial and Contemporary Ideologies of Community Management: The Case of Tank Irrigation Development in South India', *Modern Asian Studies*, 33 (2): 303–338.

Mishra, S., A. Ravindra and C. Hesse. 2013. *Rainfed Agriculture: For an Inclusive, Sustainable and Food Secure India*. London: International Institute for Environment and Development (IIED).

Nagaraj, N. and M.G. Chandrakanth. 1997. 'Intra- and Inter-Generational Equity Effects of Irrigation Well Failures – Farmers in Hard Rock Areas of India', *Economic& Political Weekly*, 13 (13): 41–44.

Namara, R.E., M.A. Hanjra, G.E. Castillo, H.M. Ravnborg, L. Smith and B. Van Koppen. 2010. 'Agricultural Water Management and Poverty Linkages', *Agricultural Water Management*, 97 (4): 520–527.

National Sample Survey Office (NSSO). 2013. *Key Indicators of Situation of Agricultural Households in India* (NSSO 70th Round, January–December 2013). Delhi: Ministry of Statistics and Programme Implementation.

National Rainfed Areas Authority (NRAA). 2012. *Prioritization of Rainfed Areas in India*. Study Report 4. New Delhi: National Innovations in Climate Resilient Agriculture (NICRA).

Raitha, S., S. Verma and N. Durga. 2014. 'Karnataka's Smart, New Solar Pump Policy for Irrigation', *Economic & Political Weekly*, XlIX (48): 10–14.

Ranganathan, T. 2014. *'Farmers' Income in India: Evidence from Secondary Data*. http://www.iegindia.org/ardl/Farmer_Incomes_Thiagu_Ranganathan.pdf. Accessed 10 March 2019.

Rao, N. 2014. 'Migration, Mobility and Changing Power Relations: Aspirations and Praxis of Bangladeshi Migrants', *Gender, Place and Culture*, 21 (7): 872–887.

Rao, N., E.T. Lawson, N. Raditloaneng, D. Solomon and M.N. Angula. 2017. 'Gendered Vulnerabilities to Climate Change: Insights from the Semi-Arid Regions of Africa and Asia', *Climate and Development*, 11 (1): 14–26.

Reddy, V. R. 2005. 'Costs of Resource Depletion Externalities: A Study of Groundwater Overexploitation in Andhra Pradesh, India', *Environment and Development Economics*, 10 (4): 533–556.

Rice, B.L. 1897. *Mysore, A Gazetteer Compiled for the Government*, Volume II. New Delhi: Asian Educational Services.

Robson, J.P. and P.K. Nayak. 2010. 'Rural Out-Migration and Resource-Dependent Communities in Mexico and India', *Population and Environment*, 32 (2): 263–284.

Rodell, M., I. Velicogna and J.S. Famiglietti. 2009. 'Satellite-Based Estimates of Groundwater Depletion in India', *Nature*, 460 (7258): 999–1002.

Saravanan, V.S., G.T. McDonald and P.P. Mollinga. 2009. 'Critical Review of Integrated Water Resources Management: Moving beyond Polarised Discourse', *Natural Resources Forum* 33 (1): 76–86.

Shah, E. 2003. *Social Designs: Tank Irrigation Technology and Agrarian Transformation in Karnataka*. Hyderabad, South India: Orient Longman.

Sharma, B.R., K.V. Rao, K.P.R. Vittal, Y.S. Ramakrishna and U. Amarasinghe. 2010. 'Estimating the Potential of Rainfed Agriculture in India: Prospects for Water Productivity Improvements', *Agricultural Water Management*, 97 (1): 23–30.

Shiva, V., H.C. Sharatchanra and J. Bandyopadhyay. 1981. *Social Economic and Ecological Impact of Social Forestry in Kolar*. Bangalore: Indian Institute of Management (IIM).

Shrestha, S., A.K. Anal, P.A. Salam and M. van der Valk. 2015. *Managing Water Resources under Climate Uncertainty: Examples from Asia, Europe, Latin America, and Australia*. Switzerland: Springer International Publishing.

Singh, C. 2018. 'Participatory Watershed Development Building Local Adaptive Capacity? Findings from a Case Study in Rajasthan, India', *Environmental Development*, 25 (197): 43–58.

Singh, C. 2019. 'Migration as a Driver of Changing Household Structures: Implications for Local Livelihoods and Adaptation', *Migration and Development*, 1–19.

Singh, C., and R. Basu. 2019. 'Moving in and out of Vulnerability: Interrogating Migration as an Adaptation Strategy along a Rural Urban Continuum in India', *The Geographical Journal*. doi:10.1111/geoj.12328.

Singh, C., A. Rahman, A. Srinivas and A. Bazaz. 2018. 'Risks and Responses in Rural India: Implications for Local Climate Change Adaptation Action', *Climate Risk Management*, 21: 52–68.

Singh, C., D. Solomon, R. Bendapudi, B.R. Kuchimanchi, S. Iyer and A. Bazaz. 2019. 'What Shapes Vulnerability and Risk Management in Semi-Arid India? Moving towards an Agenda of Sustainable Adaptation', *Environmental Development*, 30: 35–50.

Singh, C., R. Basu and A. Srinivas. 2016. 'Livelihood Vulnerability and Adaptation in Kolar District, Karnataka, India: Mapping Risks and Responses.' Adaptation at Scale in Semi-Arid Regions (ASSAR) Short Report. http://www.assar.uct.ac.za/sites/default/files/image_tool/images/138/South_Asia/Reports/Kolar. Focus Group Discussions – Short Report – July 2016.pdf. Accessed 18 July 2017.

Solomon, D. and N. Rao. 2018. 'Wells and Wellbeing: Gender Dimensions of Groundwater Dependence in South India', *Economic and Political Weekly*, 53 (17): 38–45.

Somashekara Reddy, S.T. 2007. 'Water Management the neeruganti way', in Iyengar, Vatsala *Waternama: A Collection of Traditional Practices for Water Conservation and Management in Karnataka*. Bangalore: Communication for Development and Learning.

Sri Sathyan, B.N. 1968. *Karnataka State Gazetteer: Kolar*. Bangalore: Director of Print, Stationery and Publications at the Government Press.

Thippaiah, P. 1998. *Study of Causes for the Shrinkage of Tank Irrigated Area in Karnataka*. Bangalore: Agricultural Development and Rural Transformation (ADRT) Unit, Institute for Social and Economic Change.

Truelove, Y. 2011. '(Re-)Conceptualizing Water Inequality in Delhi, India through a Feminist Political Ecology Framework', *Geoforum*, 42 (2): 143–152.

Turton, C. and J. Farrington. 1998.*Enhancing Rural Livelihoods through Participatory Watershed Development in India*. Natural Resource Perspectives no. 34. London: Overseas Development Institute.

Upadhyay, B. 2005. 'Gendered Livelihoods and Multiple Water Use in North Gujarat', *Agriculture and Human Values*, 22 (4): 411–420.

Venot, J.-P., M. Zwarteveen, M. Kuper, H. Boesveld, L. Bossenbroek, S.V.D. Kooij, J. Wanvoeke, M. Benouniche, M. Errahj, C.D. Fraiture and S. Verma. 2014. 'Beyond the Promises of Technology: A Review of the Discourses and Actors Who Make Drip Irrigation', *Irrigation and Drainage* 63 (2): 186–194.

Wilbanks, T.J. 2015. 'Putting 'Place' in a Multiscale Context: Perspectives from the Sustainability Sciences', *Environmental Science and Policy*, 53: 70–79.

Annex 4

Table 4.A.1 Rationale for choice of blocks in Kolar district

District	Block chosen	Rationale
Kolar	Kolar	Close to district headquarter
	Srinivaspura	Emphasis on horticulture and sericulture, far from district and markets
	Mulbagul	Emphasis on horticulture and sericulture, far from district and markets
	Bangarpet	Livelihood shifts from mining to others. Good infrastructure but larger shifts have impacted livelihoods. Made up of interstate migrants

Table 4.A.2 Details of villages studied in Kolar

Block	Village	Total population	% SC	% ST	Households sampled	Net % sown area	% sown area irrigated
Bangarpet	Panthanahalli	1341	53.4	0.0	29	80.5	13.1
	N.G. Hulkur	3156	43.2	0.4	51	72.5	27.3
	Hulibele	1808	28.6	1.9	42	70.3	0.7
Kolar	Kodi Ramasandra	1476	27.8	0.0	28	64.6	17.0
	Settikothanur	969	12.8	6.2	48	66.7	64.1
Mulbagul	Nangali	6359	21.1	1.8	95	38.9	65.4
	Padmaghatta	1334	14.4	2.2	38	27.3	23.9
Srinivaspura	Marasanapalli	1571	41.2	10.2	35	72.3	39.1
	Gowdathatagadde	1778	90.7	0.1	40	42.8	27.1

5 Vulnerabilities and resilience of local women towards climate change in the Indus basin

Saqib Shakeel Abbasi, Muhammad Zubair Anwar, Nusrat Habib, and Qaiser Khan

Introduction

Pakistan is a diverse country with a vast range of ecosystems, socio-economic conditions, cultures and topography. The total area of the transboundary Indus river basin is 1.12 million km,[2] distributed between Pakistan (47 per cent), India (39 per cent), China (8 per cent) and Afghanistan 6 per cent) (FAO 2012a). In Pakistan, the Indus river basin covers around 520,000 km,[2] or 65 per cent, of the country, comprising the whole of the provinces of Punjab and Khyber Pakhtunkhwa (KPK), previously called the North-West Frontier Province (NWFP), and most of the territory of Sindh province and the eastern part of Baluchistan (FAO 2012b). The Indus river basin stretches from the Himalayan mountains in the north to the dry alluvial plains of Sindh province in the south and finally flows out into the Arabian Sea. The climate of the basin varies from subtropical arid and semi-arid to temperate sub-humid in the plains of Sindh and Punjab provinces to alpine in the mountainous highlands of the north. Annual precipitation ranges from 100mm and 500 mm in the lowlands to a maximum of 2,000 mm in the mountains. Most of the precipitation is in winter and spring and originates from the west (Fowler and Archer 2006). Snowfall in the higher altitudes (above 2,500 m) accounts for most of the river run-off (Ojeh 2006). The upper Indus river basin is a high mountain region with glaciers and the mountains limit the intrusion of the monsoon.

Pakistan's is an agro-based economy, and the majority of the population depends upon agriculture for employment (Begum and Yasmeen 2011). In 2010–2011, 74.2 per cent of working women were concentrated in the agricultural sector. Most women agricultural workers were engaged in subsistence-level farming under harsh conditions and with hardly any economic security. Alongside is their unpaid work, including livestock management and vegetable farming (AGRA 2017; Government of Pakistan 2011). Climatic pressures are negatively impacting agricultural activities and increasing the vulnerabilities of women (Abubakar 2016; FAO 2015).

The Indus river basin can broadly be divided into three zones: upstream, mid-stream and downstream (ADB 2013). Women in these basins are responsible for household activities and are also engaged in crop production, livestock rearing and the cotton industry (Iftikhar et al. 2009; Kausar and Ahmad 2005).

The chapter is based on an anthropological study carried out along the Indus tributaries. It discusses women's vulnerabilities in the upper, middle and lower Indus basin, seeking to understand the differences in vulnerabilities across these three distinct geographies.

The chapter is divided into four sections. Section one reviews the literature on gender and climate-change issues with respect to the Indus river basin and tries to understand the differential vulnerabilities for women due to multiple stressors such as frequent disasters, floods and groundwater depletion in different geographical regions. Section two gives a short description of the study locations and methodology. Section three presents the perception of women across the three areas. Section four discusses these issues and concludes the chapter.

Gender and climate change in the Indus basin

Globally, women are major actors in dealing with climate shocks (IUCN 2007; UNDP 2009; UNFPA/WEDO 2009), and in particular in reducing vulnerabilities from these shocks. It is important to recognize how women in Pakistan, which ranks 12th for global vulnerability, are affected by climate change (Ahmed and Schmitz 2011: 2). There has been a noticeable increase in temperature of 0.6 degree Celsius over the last 100 years, and since 2000 a 0.08 degree Celsius increase in temperature per decade (Khan et al. 2016). The adverse effects of climatic changes are costing the economy almost US$5 billion, about 5 per cent of GDP, annually and nearly 10 million people are being displaced by climate-induced disasters (Climate Emergency Institute 2017). Climatic variability impacts people living along the Indus river, which supports the world's biggest irrigation system, and is prone to severe flooding in July and August (World Bank 2011).

Emergent changes in temperature, precipitation, the incidence of extreme events, sea level and glacial cover are expected to impact water availability, adversely affecting agricultural production, nutrition, health and household incomes. Climate variability and water availability affect marginalized women engaged in farming and livestock rearing, increasing their vulnerabilities (Abbasi et al. 2018; Najam and Hussain 2015; Nizami and Ali 2017). In the rain-fed (*barani*) areas, men have moved away from agricultural production to diversify household incomes, and more women now manage the family farm than previously. By contrast, in the canal-irrigated areas in Sindh, where agriculture is more market oriented, cotton picking offers paid work opportunities for women. In Baluchistan and Gilgit Baltistan, where socio-cultural and patriarchal norms are more binding, agriculture is virtually the sole employer for rural women (Government of Pakistan 2013).

Climate change has severe effects on the health and wellbeing of a large number of women due to their role in both productive activities and reproduction. During the 2010 floods in Pakistan, the wellbeing of 713,000 women, including 133,000 pregnant women, was affected by lack of access to freshwater, diseases, snake bites and many other health problems (UNICEF 2010: 2). Gender-differentiated roles and responsibilities, rights, access, knowledge and priorities shape vulnerabilities, often causing women to suffer disproportionately during and after disasters because of socio-economic constraints and inequalities (Brody et al. 2008; Parikh 2007). Climate change is not gender neutral: in times of crisis women and men make different decisions, and their attitudes to risk, coping strategies, adaptability, and advice-taking and information-seeking behaviours also differ (Dankelman 2008). Consequently, vulnerability and the impacts of climate change will differ for different men and women according to the intersectionality of gender with other factors and stratifications, in this case, their location (Arora-Jonsson 2011).

Among the considerable number of work tasks undertaken by women are livestock rearing, collecting fodder, cleaning animals, making dung cakes, processing animal food products such as cheese, butter and yoghurt, and even marketing them (Hamid and Afzal 2013; Iftikhar et al. 2009). Additionally, they have household responsibilities such as childcare, water collection and cooking fuel collection (Begum and Yasmeen 2011; Butt et al. 2011; Kausar and Ahmad 2005). Nevertheless, there is a common perception that women, despite being the main major producers of food, are not allowed to become landowners (FAO 2015). Cultural barriers also limit women's access to resources. The scarcity of resources induced by climate shocks increases women's workload and time poverty, intensifying their vulnerabilities (Luqman et al. 2013; Rodenberg 2009: 26).

Migration is a major adaptation strategy for climate-induced vulnerabilities but due to resource constraints, it is often the most vulnerable, including women, who cannot migrate (Gioli et al. 2014; Mueller et al. 2014). When their husbands migrate for work, these women have to construct shelters, manage the household, and look after children and the elderly (Shah 2012). In short, they have to stay put and face the impacts of climate shocks.

However, climatic variability is not always disadvantageous for women (Abbasi et al. 2017). There are important gender implications not only to the decision to migrate but also to its consequences. Cultural norms and social structures restrict women's migration. Many migrants face unsafe working environments at their destinations, exacerbating their vulnerabilities. The increased role of women in agriculture enhances their food security. Moreover, attempts at capacity building for women, for instance in the Indus area of Gilgit and Baltistan, enhance their agency (Khan and Ali 2016).

Study locations and methodology

The main aim of this study was to explore women's vulnerabilities by understanding their perceptions of climate change and to highlight their resilience in the upper, middle and downstream areas of the Indus basin. Hyndman (2001) suggests that the link between geographical location and time can be understood as "the field is here and now, not there and then". The Hunza basin, which comprises the Hunza, Nagar and part of Gilgit districts, was selected to represent the upstream area. The focus group discussions (FGDs) were carried out in Minapin, Ali-Abad and Misigar villages. Key informants for interviews were selected from the FGDs. The Soan basin, comprising Attock, Chakwal, Islamabad, Khushab and Rawalpindi districts, were selected to represent the mid-stream. There are three different rainfall zones in the Soan basin due to topographical variations. Therefore, two FGDs were conducted in each zone. The villages of Upper-Numb Romal and Dhok Chawan were selected from the high rainfall zone of the mid-stream basin (Tehsil Murree, Rawalpindi district), while Saroja and Gang were selected from the medium rainfall zone (Chakri area). Similarly, from the low rainfall zone, Akwal and Ghool villages were selected for the FGDs (Talagang area). The Chaj Doab, comprising the Chiniot, Gujrat, Jhang, Mandibahuddin and Sargodha districts, was selected in the downstream basin. FGDs were conducted in Chak-7, Bhalwal and Sadha Kambo villages of the Sarghoda district (see Figure 5.1).

This study was conducted from June to October 2017 during the "participatory assessment of climate stress, drivers and conditions leading to different levels of vulnerability" fieldwork under the Himalayan Adaptation, Water and Resilience (HI-AWARE) project.

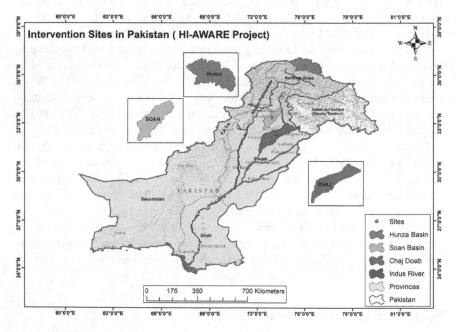

Figure 5.1 Hi-AWARE intervention sites in Pakistan.

Source: CAEWRI, NARC 2017.

Twelve FGDs from the three basin levels were conducted, each group consisting of between seven and ten young, married, widowed and elderly women. The interviews mainly focused on climate change in the area, the impacts on local livelihoods, especially those of women, their vulnerabilities and their resilience mechanisms. The information generated was analyzed to determine women's vulnerabilities and common threats in the three basins. The study sought to determine the role of culture and geography in defining gender relations, and the linkages between the changing climate and vulnerabilities.

Women's perception of climate change

Perceptions of climate change are defined as people's awareness of its causes and adverse impacts (Leiserowitz 2006; Whitmarsh 2008). These perceptions enable people to make conscious livelihood choices to address the adverse effects (Ockwell, Whitmarsh and O'Neill 2009). Public understanding is an important factor in influencing policy decisions on climate-change issues (Leiserowitz 2006). The impacts of climate change on women are serious, as they are directly related to their livelihoods (Keith et al. 2014). Women have expert knowledge of climate-change impacts and priorities at local level because of the way climate change affects their work, roles and responsibilities (Morgan 2008; Perkins 2015). However, women continue to be missing from leadership positions addressing climate-related challenges (Baruah 2016; McFarland 2015), despite the important role they are starting to adopt in adaptation strategies (Dankelman 2010). The unpaid work of women makes a major contribution to developing and implementing climate-smart adaptation practices. These often make women even more invisible in public spheres, further marginalizing them in climate-change decision-making processes (Reed et al. 2014). There is a disproportionate burden on women's unpaid labour that further increases after extreme climatic events (McGregor 2012; Parikh and Denton 2003) (Table 5.1).

Impact of climate change on women in the upstream basin

The FGDs in the upstream basin sought to understand women's perceptions of climate change-related vulnerabilities and their local adaptation strategies. The majority of women noted an increased incidence of flooding, with 65 per cent reporting greater frequency of flash floods in recent years. A significant proportion of the households also reported that they observed erratic rainfall patterns, with an increased number of landslides in summer and a reduction in snowfall.

Women form a very significant proportion of the workforce in food production and have an important role in the sustainability of the mountain environment and its cultivation (FAO 2015). Due to the emigration of their menfolk, women have to deal with all day-to-day activities. In the upstream basin, agriculture has been the major source of livelihood for the people,

Table 5.1 Socio-economic situation of women in Indus

	Literacy Level	Mobility	Social Barriers	Household Work	Farm Work	Employment	Health	Recreational Activities
Up-Stream	High (depending upon family income but overall culture for women education is very positive)	Easy, flexible and more comfortable	Strict social barriers in case of mixing with other communities and sects but somehow relaxed in terms of decision making	Fully Responsible	Equal Participation with men	No Social or Cultural Barriers	Basic health care facilities are available in the area but for complicated cases, women have to travel long to reach Gilgit City. Becomes vulnerable during extreme climatic events due to landslides and travel disconnections	Available at community and educational institutions. Higher participation rate no family restrictions
Mid-Stream	Medium (But the trend of women education is increasing with passage of time as compared to past)	Easy within the village but for outside areas neither flexible nor comfortable without accompany of male family member	Doesn't have right to make decisions or take part in any discussion	Fully Responsible	Participation rate is high but lower as compared to Up-Stream Basin	Allowed at limited scale and professions such as teaching and Health care	Basic health care facilities are available in the area but for complicated cases, women have to travel long distances, mostly towards city area of Rawalpindi and Islamabad	Available at educational institutions and within household levels. Mostly the recreational activities are in shape of festivals. Participation in sports etc is not encouraged

| Down-Stream | Low (Still on lower side as compared to other two basins because of social and cultural barriers) | Only allowed within community or village. Women are not allowed to go out without male member. Mobility is strict | No rights have assigned to women she is not allowed to interfere in any decision taken by male member | Fully Responsible | No participation in crop and livestock activities. Only poor households or tenants women work in farm | Mostly strict but teaching is allowed in near areas. | Health care facilities are available at government and private level and also at nearby areas. Lack of facilities at government hospitals create certain issues. | Limited only at household level, during marriages and other religious festivals. |

Source: SSRI and NARC survey data.

with 70 per cent of the population dependent on it directly or indirectly. The emigration rate in Hunza district was the highest in Gilgit Baltistan, resulting in a labour shortage for agriculture (Gioli et al. 2014). From the FGDs it was observed that this trend is negatively impacting agricultural productivity. The main staple crop is wheat, but maize and barley are also cultivated. Cultivating fruit like apricots, peaches, pears, apples, grapes, cherries and melons, and vegetables such as potatoes, tomatoes and beans is an important source of income for local people. Women play a major role in the upstream basin, being solely responsible for value addition. Climatic changes, they said, have led to low agricultural production and farm incomes, particularly for their fruit crops. A substantial proportion of households reported a decline in the production of almost all horticultural crops.

The majority of the women from Gulmit village (upstream) mentioned in discussions that flash floods and landslides caused by climate change are becoming more common in the area and these occurred almost every year (see Figure 5.2). However, they also said that vulnerabilities affect women more than men because men are better equipped to act in a particular challenging situation than women. They added that men have better opportunities for mobility while women remain at home, further contributing to their level of vulnerability.

As well as hazards, households face food insecurity as their local livelihood sources are destroyed. Consequently, households' dependency on external food items supplied from downstream areas has increased. Furthermore, the food supply from other areas also gets interrupted due to infrastructure damage in the hazard-hit area. Vulnerabilities in the shape of food insecurity, poor infrastructure and physical isolation increase, and rehabilitation requires physical and financial resources.

The frequent flood events of the last decade have resulted in health issues for women, primarily due to poor sanitation. Women in the area had to sacrifice their own needs, including food, to meet their family's needs. Diseases, infections and injuries are more common in women during floods

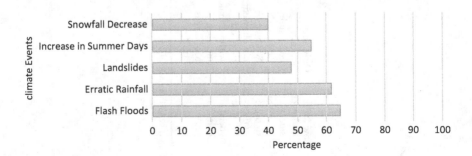

Figure 5.2 Women's perception: Climate induced events in UpStream Basin.
Source: SSRI, NARC survey data, 2017.

when they become homeless, or are living with relatives or in tents. Another important factor increasing climate-induced vulnerabilities for women in the upper Indus is the out-migration of men from the hazard-hit areas. With the men away, the women have to undertake men's tasks at home. Availability of clean drinking water after the floods is another major problem that upstream communities confront. Fetching water for domestic consumption is women's responsibility. They have to travel long distances in the harsh mountain areas. This is not only dangerous but also results in declining health, especially for pregnant and lactating women.

Women felt that in the upstream basin, climate change has led to significant degradation of open pastures and rangelands. Since sheep and the larger animals consume more fodder and water, as a coping strategy the women said they have reduced the number of their cattle and small ruminants. During environmental shocks such as floods, prolonged dry spells, drought or erratic rainfall, most farming households face food insecurity due to damage to their farming systems and other livelihood sources. The FGD participants mentioned that they had faced at least one shock in the preceding 12 months. The women mentioned that during these events households adopt various strategies to reduce the impact of shocks, such as replacing expensive food items with cheaper ones or borrowing from relatives. Most of them said they had reduced expenditure on clothing.

Some women mentioned that they changed their farming practices, growing short-duration crops to enhance their resilience to climate change. New crops, including seasonal vegetables, nuts and hybrid potatoes that withstand water stress and have a higher market value, are being grown with the help of NGOs. New farming practices include changing the sowing time and water conservation.

Women's perception of climate change in the mid-stream basin

The southern part of the mid-stream basin gets very hot in summer, while the northern parts experience very cold winters with heavy snowfall. January is usually the coldest and June the hottest month in the year. The area is at risk from erratic rainfall, hailstorms, droughts, heatwaves and cold waves. Crop cultivation and rearing livestock are becoming difficult for small farmers.

The FGDs showed that women were actively participating in agricultural activities and livestock rearing. Over the last three decades, women's living conditions have improved due to an increase in income, education and awareness through mass media. They are involved in important family decisions but still do not have the right to decide about marriage. Gender roles and responsibilities are fixed. Women are primarily responsible for domestic activities and men mainly perform outdoor activities. Women also contribute to farming activities ranging from land preparation to harvesting. In livestock management, women perform 70–80 per cent of the tasks.

Women's health is affected by extreme heat during wheat harvesting. They reporting that collecting water from springs was a tough job and they often suffered from neck pain carrying water over long distances on their heads. Mothers expressed safety concerns about their daughters fetching water from a distance.

A woman from Saroba village pointed out during discussion that climate change-induced drought and heat stress is causing their agricultural land to become barren, increasing their poverty level. She mentioned that her husband is a daily labourer and during extreme weather his income is halved. She further added that male members can migrate but women do not tend to do so due to climate change, which increases their vulnerability level compared with their male counterparts. This woman also highlighted that due heat stress, women have more skin problems than men, as women left behind in the village are responsible for agriculture and livestock while the men migrated to the nearest city.

Rainfall variability further increases women's workload. Lack of rain causes seasonal scarcities in most parts of the mid-stream basin, increasing the burden on women who have to fetch water over long distances. Poor sanitation and hygiene practices resulting in water-borne diseases are found more commonly in women than men. Climatic variability in cropping patterns has also affected women. As men are responsible for buying seeds and sowing crops, their farming responsibilities end before those of the women. Women suffer more as they are engaged on farm work right up till harvesting, in addition to livestock and household management.

In the mid-stream basin livestock is a major source of livelihood and everyday routines are different from those in the upstream and downstream basins. In upstream areas, livestock is grazed on a collective community system. A few young men take the livestock of the entire community to the high-altitude pastures and live there for the whole summer. Individual households do not have to tend their animals daily in summer, and in winter, there is enough dry fodder for them. In downstream areas, livestock is kept in sheds in the farms and managed primarily by men. In the midstream areas, however, taking livestock to graze on nearby land is a daily practice.

High rainfall zone (Tehsil Murree)

Results from the FGDs indicate that in most households, the men are away from home, working in other cities, suggesting a larger burden of responsibility for women running the household, farmland and other village matters. Women are traditionally responsible for cooking, fetching drinking water, cleaning, feeding and milking livestock, sowing, weeding, harvesting, home gardening and so on. In Tehsil Murree, women were asked to recall the impact of climate change over a decade. They recalled that 10 years previously there had been green and fertile fields, where multiple crops including fruit and vegetables were grown. Livestock was kept by every family and water was easily available. But now everything has changed. The

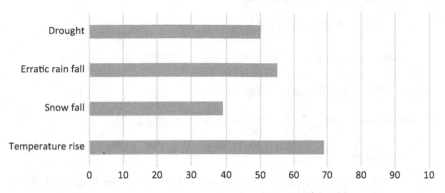

Figure 5.3 Women's perception: Climate-induced events in the High Rainfall Zone MIB (Pakistan).

Source: SSRI, NARC survey data, 2017.

orchards are barren and cereals are hardly cultivated. Few households still grow vegetables for their own consumption as well as a little for the market.

The women felt that winter snowfall had reduced, leading to livelihood vulnerabilities in the area. as crop yields were affected due to lower soil moisture. The majority of the women felt the temperature was rising. Erratic rainfall and drought were other climatic stressors reported. Women also mentioned snowfall as a climatic vulnerability. A number of health issues, including common cold, fever, throat infections and skin issues, were reported due to cold dry weather. These health issues are more common in women and children – in women due to more exposure to cold weather early in the morning, late nights and water-related activities (see Figure 5.3).

Medium rainfall zone (Chakri)

Women from Chakri and Sarooba villages, who rely on natural resources for their livelihood and food security, said that climate change is part of their everyday life. The women mentioned that droughts are affecting their socio-economic situation. They mentioned being susceptible to rising temperatures, erratic rainfall, droughts and fog (Figure 5.4). Uncertain rainfall made agriculture unreliable. Migration is seen as a major adaptation strategy to cope with climatic stressors.

Low rainfall zone (Tehsil Talagang)

The women in Ghool and Akwal villages spoke about extreme heat, changing weather patterns and how in the past they had been able to depend on the regular rains. Nowadays they said the rains were irregular and accompanied by high winds. The women talked about the early rains in summer and said that last year, heavy rains accompanied by high winds and hailstorms had

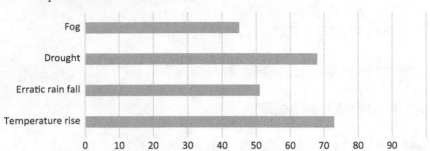

Figure 5.4 Women's perception: Climate-induced events in the Medium Rainfall
        Zone MIB (Pakistan).

Source: SSRI, NARC survey data, 2017.

damaged standing wheat and pulse crops. According to them, the intensity
of both summer heat and winter cold had increased. These women perceive
climate in terms of weather. To them climate meant winter, summer, heat,
cold, rainfall, drought – the weather they experienced and lived with every
day. Compared to a decade ago, seasonality and occurrences are changing,
and the weather is unpredictable. A woman from Ghool village said, "Now
the summers are longer and hotter. It is difficult to work in the intense heat.
We have no electricity, and there is a severe water shortage. We feel sick,
tired and uncomfortable, but we have to carry on with our daily domestic
work." Other women added that the intense heat does not allow them to
stay outdoors for long. Winters are colder than in the past. People begin
work in the fields later in the day when it gets somewhat warmer. These
adjustments to lifestyles and work schedules represent people's response to
weather changes. In these two villages, the majority of women felt temperatures had risen and erratic rainfall patterns and droughts had increased in
the last decade; they also identified fog as a new phenomenon (Figure 5.5).

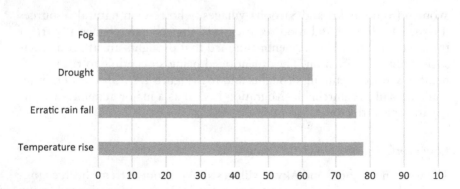

Figure 5.5 Women's perception: Climate-induced events in the Low Rainfall Zone
        MIB (Pakistan).

Source: SSRI, NARC survey data, 2017.

Women's perception of climate change in the downstream basin

In the lower basin area, women's participation in agriculture was very limited, due to social norms that restrict them from working outside the home. Highly mechanized commercial agriculture further restricted women's participation (Figure 5.6). Large farms located some distance from the village do not create a suitable working environment for women. This is in contrast to the upstream and mid-stream areas, where subsistence farming in small plots located in the villages enable women to work in them. Sugar cane and wheat are the major crops and citrus fruits provide a major source of income.

Women here are not much exposed to farming and have limited interest in agriculture due to their higher level of education. Here both men and women emigrate for jobs or to seek higher education. Households with migrants have better-constructed houses and household assets, and better access to education and health facilities. However, the migration of men increases the workload of the women left behind. Overall, in around 10 per cent of families in the area, one or more men have migrated out of the village in search of better opportunities.

Women felt that over the last 10 to 20 years, the rains have become erratic, but annual precipitation has decreased. Rainfall is more intense but of shorter duration, followed by long dry spells. There is heavy fog in winter. Summers are longer, and winters are shorter. Changes in temperature and rainfall have also aggravated pest attacks on crops and livestock, threatening the livelihoods of the rural poor in the downstream area. A girl in village Chak-7 highlighted one very basic issue. She said that although climate change strikes both genders equally, when it comes to adaptation in the face of vulnerabilities women's freedom to follow their men is restricted because of their limitations in terms of mobility and access to resources. She stated that at times of climate-induced vulnerability such as flooding,

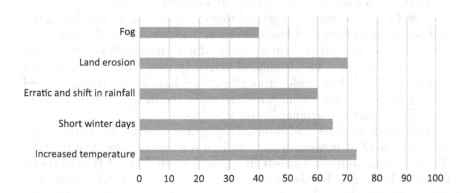

Figure 5.6 Women's perception: Climate-induced events in the Low Indus Basin LIB (Pakistan).

Source: SSRI, NARC survey data, 2017.

women sacrificed their savings and even their jewellery. She added that girls are not allowed to keep mobile phones with them, but sometimes men have more than two phones.

Older women said that over time the condition of women has improved due to the increased incomes, improved education and education through mass media. Women were involved in important family decisions but had no say regarding marriages. In flood-prone areas of the basin, women said that some farmers have started cultivating sugar cane instead of rice as it is more resilient to floodwaters. Additionally, if their requirements cannot be fulfilled from canals, some farmers, especially citrus growers, are adopting improved irrigation techniques such as sprinklers and drip irrigation.

The FGDs revealed that though flooding was not common in the area, the floods of 1992 and 2010 caused a lot of damage. Changing patterns of rainfall, with heavy rains occurring just before the wheat harvests, the women said, caused most farmers to lose 100 per cent of their standing crop. Communities living in waterlogged areas are more vulnerable to crop losses. Most women said that during disasters they suffer the most due to limited access to financial, natural, institutional or social resources and also social norms like dress codes that hinder mobility. On the whole, the majority of women felt temperatures were rising, and had experienced land erosion, shorter winters, erratic rainfall and fog.

Learning from women's vulnerabilities and responses to the changing climate

This section draws on the vulnerabilities and capacities of women in the three basins of the Indus. (Table.5.2). It is apparent that climate change resulting in floods, changing rain cycles, river erosion, flood, heatwaves and cold impact human capital. It is, however, also clear that both women and men have the agency to meet the outcomes of climate shocks.

In the FGDs, the women explained that changing rain cycles, floods, soil and riverbank erosion and landslides, and heatwaves were the aspects of climate change that affect them the most. Major catastrophes deprive them of their lands, homes and livelihood assets. During these events, women undertake all the household tasks while men have to build shelters and provide food. However, women have to make the shelters liveable, and to cook, fetch water and arrange livestock fodder.

With harsh weather and difficult topography in the upstream basin, the daily life of men, women and children is precarious most of the time due to climate-induced hazards like glacial melts, glacial lake outburst floods (GLOF) and floods. Despite being custodians of culture, women are the most vulnerable in the mountains and must work harder than men to adapt.

In the mid-stream areas, women's drudgery is compounded by social barriers. They are not involved in major decisions and do not have land

Table 5.2 Women's vulnerabilities to climate change in Indus basin

Basin	Vulnerabilities	Threats and Stressors	Social and Cultural Norms	Coping and Adaptation Strategies
Up-Stream	– Harsh weather – Difficult topography – Climate induced hazards – Lack of capacity building – Lack of opportunities – Small agricultural lands	– Glacial melt – GLOFS – Floods – Outmigration	– Women are custodian of household farming – Women doesn't have land ownership – Women are not allowed to participate in decision making	– Out-migration – Cropping pattern changed according to weather
Mid-stream	– Small landholdings – Less resources – Women health issues – Gender biased society – Climatic variability	– Fog – Drought – Heat and cold waves – Erratic/unpredictable rainfall	– Involved in agricultural and livestock activities – Responsible for household activities/childcare	– Out-migration – Involved in private and public employment
Down-stream	– Degradation of natural resources – Less availability of resources for everyday life – Water logging & salinity – Social norms	– Climatic variabilities – Heavy Fog – Erratic rainfall – Droughts	– Women restricted to home – Women are excluded from education – Women are excluded from employment – Women are not allowed to move out without family head	– Out- migration – Education/awareness programs for local women

Source: SSRI, NARC survey data.

ownership rights. Though the Islamic Shariah law define men's and women's shares in property, in practice these stipulations are ignored or waived (Ahmad 2010; SDPI 2011). Women generally have limited resources for subsistence. However, they are equally responsible for the care of livestock.

Rural customs and urban norms reinforce gender inequality in the midstream basin. Women are far behind men here. One of the reasons for this is their low representation in governance. On the other hand, climate change is adversely impacting agriculture, water and food security severely, affecting their health. Over the last decade, temperature and precipitation patterns have significantly changed, increasing the vulnerability of women and children.

In the downstream villages, due to commercialization and mechanization of agriculture, women are kept away from agriculture and are responsible for household chores. Social barriers and norms do not allow women to go out without male family members in contrast to the upstream areas where women have freedom of movement. In the mid-stream areas, there are no restrictions on women accessing education, but they are not encouraged to move out of the city and or to find work.

Climate, and in particular rising temperatures, is negatively impacting agriculture, exacerbating the existing socio-economic stresses of the many factors impacting livelihoods, such as lack of medical facilities for both humans and livestock, and the unavailability of pesticides. Furthermore, lower-income groups who lack savings are forced to borrow from informal credit sources at high rates of interest. Sometimes they have to sell their livestock, or in extreme cases, to sell some land for less than its true value.

In terms of adaptation practices against vulnerabilities, women have to follow their men due to financial and social constraints. Men dominate the labour markets; women's agricultural work is mostly unpaid. An emerging phenomenon is the presence in the labour market of women from poorer households. This has improved their lives and the adaptation capacities of their households. But social barriers restrict mobility for the majority of the women. Saving money is the most significant adaptation strategy for women in the mid- and lower basins; almost all the women strongly agreed that this enabled them to meet contingencies like weddings, deaths, children's education and floods. While these savings are not sufficient to meet all costs in crisis situations, they are enough to meet the most important ones. In most countries of South Asia, women save money for emergencies either through cooperatives or individually, hiding it from the men who might spend it. The other important common adaptation strategy for women is rearing small ruminants that can be sold when needed. Women are responsible for fodder for livestock and poultry. They said they save dry fodder against disasters and in advance of the rains.

Women in the mid-basin have access to health facilities. Those in the higher and lower basins do not, so they use home remedies. During heat stress, women adapt by avoiding unnecessary outdoor activities and using traditional drinks.

Due to the increasing migration of their men, women manage the household, farm and livestock. At times of extreme climate events and disasters, their workload greatly increases, often at the cost of their health. Nevertheless, women are rapidly adapting to climate change thanks to the local NGOs. New houses and animal sheds are being constructed in safer places; crop rotations and climate-smart agricultural practices have been introduced in the area. Women are learning new skills for value addition to crops to enhance their incomes, and buying smaller livestock to save scarce water resources. They are increasing their savings in local cooperatives.

This chapter has highlighted how climate change has enhanced both the vulnerability and the agency of women. Although women in Pakistan have the right to land by law, they rarely own it, and in some downstream areas culture prevents women from participating in agricultural activities. Although overall, women's link to agriculture is strong, as it provides food security, this relationship remains location specific. Upstream, about 70 per cent of women work in agriculture due to a shortage of labour. Downstream, women do not work in agriculture as it is being mechanized so the work is moving beyond their reach. Some women in mid-stream areas are educated and choose not to work in agriculture, opting instead to cultivate fruit and vegetables, which are relatively more resilient to water stress and command a higher market value. In some places sugar cane, a cash crop, has replaced rice as it is more resilient to floodwaters. In the canal region, as water decreases there is a change to sprinklers and drip irrigation.

Climate variability contributes to the vulnerability of women when there is snowfall, fog or heat. Research shows that this is a major problem impacting the vulnerability of women, who are finding it difficult to work outside for longer with summers getting longer. For instance, women in Murree are affected by the heat. According to the research, "Women differ from men in their physiologic compensation to elevated temperatures, which contributes to their biologic vulnerability"; other factors include "access to healthcare and cooling facilities due to personal safety concerns and a lack of access to personal transportation, culturally prescribed heavy clothing garments that limit evaporative cooling, and a lack of awareness of women's vulnerabilities to heat" (Sorenson et al. 2018: 284). Women may need to learn further to adapt to heat and climate change; one suggestion is to train them to protect themselves from heat waves (Abubakar 2016). The distinct geographies and multiple stressors highlighted in this chapter, such as water, disasters, snowfall and heat, play a role in a cultural shift that is seeing women step out of their earlier roles and adapt to the vulnerabilities inflicted by climate change.

Women do not remain without agency; they adapt to change when there is a need. Overall, this study shows that women are adaptive and are capable of coping with the adverse impacts of climate change, despite severe social and cultural constraints on their mobility and decision making on issues such as marriage. Women can therefore play an important role in building resilience if their voices are raised and heard. It is clear that there is an urgent need to involve women in change-making climate adaptation strategies and policies. Capacity building through training on disaster risk reduction (DRR), and climate-smart interventions such as rainwater harvesting and tunnel farming would increase resilience. Adaptation is the key to removing vulnerability and increasing resilience, and this study has shown women's ability to adapt in various sectors. Access to, and control and ownership by women of the resources of land, livestock, property and income opportunities must be recognized by policymakers and wider society if this process is to be further enhanced.

References

Abbasi, S., Muhammad, A., Nusrat, H., Kaiser, K. and Kanwal W. 2018. 'Identifying Gender Vulnerabilities in Context of Climate Change in Indus Basin', *Environmental Development*, 1: 34–42.

Abbasi, S.S., Ahmad, B., Ali, M., Anwar, M.Z., Dahri, Z.H., Habib, N., Hussain, A., Iqbal, B., Ishaq, S., Mustafa, N., Naz, R., Virk, Z.T. and Wester, P. 2017. 'The Indus Basin: A Glacier-fed Lifeline for Pakistan'. HI-AWARE Working Paper 11. Kathmandu: HI-AWARE.

Abubakar, S.M. 2016. 'Women and Climate Change'. *The International News.* https://www.thenews.com.pk/print/116402-Women-and-climate-change. Accessed on 28 September 2019.

ADB. 2013. '*Indus Basin Floods: Mechanisms, Impacts, and Management Mandaluyong City, Philippines*'. Mandaluyong: Asian Development Bank.

AGRA. 2017. '*Africa Agriculture Status Report: The Business of Smallholder Agriculture in Sub-Saharan Africa.*' Nairobi: Alliance for a Green Revolution in Africa (AGRA), Issue No. 5.

Ahmad, N. 2010. 'Land Rights for Pakistani (Muslim) Women: Law and Policy 1.' Policy Brief Series No. 23. SDPI Islamabad. https://sdpi.org/publications/files/Microsoft%20Word%20-%20policy%20Brief%2023.pdf. Accessed on 1 December 2019.

Ahmed, N.M. and Schmitz, M. 2011. 'Economic Assessment of the Impact of Climate Change on the Agriculture of Pakistan', *Business and Economic Horizons*, 1–12.

Arora-Jonsson, S. 2011. 'Virtue and Vulnerability: Discourses on Women, Gender and Climate Change', *Global Environmental Change*, 21: 744–751.

Baruah, B. 2016. 'Renewable Inequity? Women's Employment in Clean Energy in Industrialized, Emerging Developing Economies', *Natural Resources Forum. A United Nations Journal*, 41(1): 18–29.

Begum, R. and Yasmeen, G. 2011. 'Contribution of Pakistani Women in Agriculture: Productivity and Constraints', *Sarhad Journal of Agriculture*, 27(4): 637–643.

Brody, A., Justin, D. and Emily, E. 2008. '*Gender and Climate Change: Mapping the Linkages. A Scoping Study on Knowledge and Gaps.*' BRIDGE, Institute of Development Studies (IDS), UK.

Butt, T.M., Hassan, Y.Z., Mehmood, K. and Muhammad, S. 2011. 'Role of Rural Women in Agricultural Development and their constraints', *Journal of Agricultural Science*, 6(3): 53–56.

Climate Emergency Institute. 2017. 'Global Warming and its Impacts on Pakistan.' Retrieved from https://www.climateemergencyinstitute.com/uploads/global_warming_and_its_impacts_o n_Pakistan.pdf. Accessed on 4 August 2019.

Dankelman, I. 2008. '*The Issues at Stake: Gender, Climate Change and Human Security.*' *Women's Environment and Development Organisation, cited in Women's Health Victoria, 10-Point Plan for Victorian Women's Health 2010– 2014, 2010.*

Dankelman, I. 2010. *Gender and Climate Change: An Introduction.* Routledge.

FAO. 2012a. '*Irrigation in Southern and Eastern Asia in figures.*' AQUASTAT Survey 2011. Rome: FAO, 129–140.

FAO. 2012b. AQUASTAT Database, Dataset for Country Water Resources Utilized. Retrieved from http://www.fao.org/nr/water/aquastat/main/index.stm. Accessed on 21 September 2019.

FAO. 2015. 'Women in Agriculture in Pakistan.' Retrieved from http://www.fao. org/3/a-i4330e.pdfFSAf. Accessed on 9 October 2019.

Fowler, H.J. and Archer, D.R. 2006. 'Conflicting signals of climate change in the Upper Indus basin', *Journal of Climate*, 19: 4276–4293.

Gioli, G., Khan, T., Scheffran, J., Aneel S, Haroon, U.T., Niazi, I. 2014. 'Gender and Environmentally-induced Migration in Gilgit-Baltistan, Pakistan.' In, *Sustainable Development in South Asia: Shaping the Future.* Islamabad: Sustainable Development Policy Institute and Sang-e-Meel Publishers, 355–378.

Government of Pakistan. 2011. *Labour Force Survey 2010–11* (4th Quarter). Islamabad: Statistics Division, Pakistan Bureau of Statistics.

Government of Pakistan. 2013. *Labour Force Survey 2012–13.* Islamabad: Statistics Division, Pakistan Bureau of Statistics.

Hamid, A.J. and Afzal, J. 2013. '*Gender, Water and Climate Change: The Case of Pakistan.*' Pakistan Water Partnership Policy Paper Series.

Hyndman, J. 2001. *Geographical Review.* 19(1/2): 262–272.

Iftikhar, N., Ali, T., Ahmed, M., Maan, A.A. and Haq, Q. 2009. 'Training Needs of Rural Women in Agriculture: A Case Study of District Bhawalpur.' *Pakistan Journal of Agricultural Science* 46 (3): 200–208.

IUCN. 2007. Gender and Climate Change: Women as Agents of Change. Retrieved from http://cmsdata.iucn.org/downloads/climate_change_gender.pdf. Accessed on 11 December 2019.

Kausar, T. and Ahmad, S. 2005. 'Social Stratification in the Participation of Women in Agricultural Activity: A Case Study of District Khushab Punjab, Pakistan', *Pakistan Geographical Review*, 60(2): 80–86.

Keith, P., Sibonokuhle, N. and Tanyaradzwa, B.C. 2014. Climate Change Impacts on Rural Based Women: Emerging Evidence on Coping and Adaptation Strategies in Tsholotsho, MCSER Publishing, Rome-Italy. *Mediterranean Journal of Social Sciences*, 5(23): 2545. doi:10.5901/mjss.2014.v5n23p2545

Khan, A.M., Khan, A.J., Ali, Z., Ahmad, I. and Ahmad, N.M. 2016. 'The Challenge of Climate Change and Policy Response in Pakistan', *Environment Earth Science*, 27: 412.

Khan, M. and Ali, Q. 2016. 'Socio-Economic Empowerment of Women in Pakistan; Evidences from Gilgit-Baltistan', *International Journal of Asian Social Science, Asian Economic and Social Society*, 6(8): 462–471.

Leiserowitz, A. 2006. 'Climate Change Risk Perception and Policy Preferences: The Role of Affect, Imagery, and Values', *Climatic Change*, 77(1–2), 45–72. doi:10.1007/s10584-006-9059-9.

Luqman, M., Shahbaz, B., Khan, I.A. and Safdar, U. 2013. 'Training Needs Assessment of Rural Women in Livestock Management: Case of Southern Pujab, Pakistan', *Journal of Agriculture Research*, 51(1): 99–108.

McFarland, J. 2015. 'Are there jobs for women in green job creation?' *Women & Environments, 89/90, fall 2014/winter 2015*. Retrieved from www.weimagazine. com.

McGregor, D. 2012. 'Traditional Knowledge. Considerations for Protecting Water in Ontario', *International Indigenous Policy Journal*, 3(3), Art. 11. Retrieved from https://ir.lib.uwo.ca/iipj/vol3/iss3/11.

Morgan, C. 2008. 'The Arctic: Gender issues', *Parliamentary Information and Research Service Publication PRB*, 08–09E.

Mueller, V., Gray, C. and Kosec, C.. 2014. 'Heat stress increases long term migration in rural Pakistan', *National Climate Change*, 4(3): 182–185.

Najam, N. and Hussain, S. 2015. 'Gender and Mental Health in Gilgit-Baltistan, Pakistan', *Journal of Pioneering Medical Sciences*, 5: 117–120.

Nizami, A. and Ali, J.. 2017. 'Climate Change and Women's Place-based Vulnerabilities – a Case Study from Pakistani Highlands', *Climate and Development*, 9(7): 662–670. doi:10.1080/17565529.2017.1318742.

Ockwell, D., Whitmarsh, L. and O'Neill, S. 2009. 'Reorienting Climate Change Communication for Effective Mitigation Forcing People to be Green or Fostering Grass-Roots Engagement?', *Science Communication*, 30(3): 305–327. 10.1177/1075547008328969.

Ojeh, E. 2006. Hydrology of the Indus Basin. Retrieved from https://webspaceu-texas.edu/eno75/HYDROLOGY%20OF%20THE%20INDUS%20BASIN%20 by%20Elizabeth%20Ojeh.doc. Accessed on 10 February 2019.

Parikh, J. 2007. 'Gender and Climate Change: Framework for Analysis, Policy and Action. Report by UNDP India.' Retrieved fromwww.data.undp.org.in/Gender_CC.pdf. Accessed on 18 June 2019.

Parikh, J. and Denton, F. 2003. 'Gender and Climate Change', at COP8: A forgotten element.

PBS. 2019. Paksitan Bureau of Statistics. Retrieved from http://www.pbs.gov.pk/ sites/default/files/population_census/Administrative%20Units.pdf. Accessed on 12 June 2019.

Perkins, P.E. 2015. 'Gender and Climate Justice in Canada', *Women and Environments. International Magazine*. (Special Issue: Women and Work in Warming World), 94/95, fall 2014/Winter 2015, 17–20.

Reed, M.G., Scott, A., Natcher, D. and Johnson, M. 2014. 'Linking Gender, Climate Change, Adaptive Capacity, and Forest-Based Communities in Canada', *Canadian Journal of Forest Research*, 44(9), 995–1004.

Rodenberg, B. 2009. '*Climate Change Adaptation from a Gender Perspective*'. Bonn: German Development Institute.

SDPI. 2011. Women and Land. In-focus case study. Islamabad IDRC. CRDI. https:// www.idrc.ca/sites/default/files/sp/Documents%20EN/Pakistan-in-focus-case-study-women.pdf. Accessed on 24 December 2019.

Shah, A.S. 2012. 'Gender and building homes in disaster in Sindh, Pakistan', *Gender & Development*, 20(2): 249–264.

Sorenson, C.S., Saunik, M., Sehgal, A., Tewari, M., Govindan, J.L. and J. Balbus. 2018. 'Climate Change and Women's Health: Impacts and Opportunities in India.' *Advancing Earth and Space Science*. https://doi.org/10.1029/2018GH000163. Accessed on 21 February 2019.

UNDP. 2009. 'Resource Guide on Gender and Climate Change.' Retrieved from http://content.undp.org/go/cmsservice/download/asset/?asset_id=1854911. Accessed on 24 September 2019.

UNFPA/WEDO. 2009. Climate Change Connections. Gender and Population. Retrieved from http://www.wedo.org/category/act/climate-change-toolkit. Accessed on 27 September 2019.

UNICEF. 2010. 'Pakistan Monsoon Floods Immediate Needs for Women and Children affected by Monsoon Floods', 5 August 2010, Pakistan. https://www. unicef.org/french/infobycountry/files/UNICE__Immediate_Needs_Document_ for_Pakistan_5_August_2010.pdf. Accessed on 25 December 2019.

Whitmarsh, L. 2008. 'What's in a name? Commonalities and differences in public understanding of "climate change" and "global warming"', *Public Understanding of Science*, 18(4): 401–420. doi:10.1177/0963662506073088 (21 October 2019).

World Bank. 2011. *Vulnerability, Risk Reduction and Adaptation to Climate Change*. Washington, DC: Climate Risk and Country Profile Pakistan.

6 Climate change, gendered vulnerabilities and resilience in high mountain communities

The case of Upper Rasuwa in Gandaki River Basin, Hindu Kush Himalayas

Deepak Dorje Tamang and Pranita Bhushan Udas

Introduction

There is growing recognition of the need for context-specific understanding of gendered vulnerabilities to address issues of climate change and consequent socio-economic changes (Goodrich et al. 2019a). Mountains are climate 'hot spots' (Khan et al. 2018), and communities living in rugged and difficult mountain terrain are more vulnerable to climate change than those living in the plains (Sharma et al. 2019). Mountain people contribute significantly to the lives of downstream populations by providing, among other things, rare high-value medicinal resources. Yet their wellbeing indicators show how vulnerable they are to a changing climate (Gerlitz et al. 2011; Humagain and Shrestha 2009). Basic facilities like water, energy, health care and education are scarce in the high mountains. Limited public services provided by the government are difficult to access due to lack of capabilities (Pasteur 2011). Climatic stressors have further increased the vulnerabilities of the mountain communities and intensified existing disparities. Understanding and responding to their vulnerabilities is crucial not only for climate change adaptation initiatives but also for achievement of the Sustainable Development Goals.

This chapter examines gendered vulnerabilities in the high mountain villages of the upper Rasuwa district of Nepal.[1] This is part of a region popularly known as the Hindu Kush Himalayas. People living in the district are witnessing the changing climate and its impact on their day-to-day lives (Campbell 2017). The study seeks to understand people's livelihoods in the context of climate change in a scenario of high social and gender differentiation. In this chapter, we examine how people belonging to different genders, castes, classes, ethnic, education and age groups are coping with the changes. This analysis is expected to contribute to gender-sensitive adaptation planning for the high mountain regions of Nepal.

The chapter is organized in six sections. Following this introduction, we discuss research methodology and introduce the study villages. Section three conceptualizes and highlights gendered vulnerability in the high mountain regions. In the fourth section we analyze the gender roles and responsibilities across the various livelihood strategies undertaken by differently positioned households. Based on evidence from section four, section five analyzes gender vulnerabilities. The last section presents conclusions from the analysis.

Research methodology and study area

This chapter is based on a larger study of gendered vulnerabilities in the Nepalese and Indian part of the Gandaki river basin (Figure 6.1) under the Himalayan Adaptation, Water and Resilience (HI-AWARE) research project undertaken by the International Centre for Integrated Mountain Development between 2015 and 2018.[2] The chapter synthesizes findings from upstream areas of the basin, with cases from Nepal's Rasuwa district, home to about 89,000 people. The study villages are located in the northern part of the district.

The study uses qualitative methods of data collection. Two sets of primary data are analyzed: data collected in 2016 that identifies drivers and conditions of socio-economic vulnerabilities; and data collected in 2017 on gender vulnerabilities. Transect walks, participant observation, key informant interviews, focus group discussions (FGDs) and case studies were used. For the gender vulnerability analysis, 40 FGDs were conducted with homogeneous groups of women and men belonging to similar caste/ ethnic groups, of similar socio-economic status as reflected in their food sufficiency (more than 9 months, up to 6 months and up to 3 months) and using similar livelihood strategies. A semi-structured checklist was used to obtain inter-generational and intersectional information from the respondents.

The mountainous upstream section of the basin has low population density due to the hardships of steep slopes, snow and lower land fertility. The average population density of the upstream districts in Nepal – Mustang, Manang and Rasuwa – is nine persons per square kilometre, compared to 183 and 278 persons per square kilometre respectively in the mid- and downstream districts of the basin (Dandekhya et al. 2017).

Increased temperature and changing precipitation patterns are observed in upstream areas. The HI-AWARE study on 30-year temperature data (1981–2010) show that both maximum and minimum temperatures have increased by nearly 1.55 °C and 0.0521 °C in upstream areas in all seasons, with the highest trend in the post-monsoon season (all significant at the 95 per cent confidence level). Annual precipitation has decreased by 59 mm in the last 30 years, with the winter dry period increasing significantly (HI-AWARE 2017).

Another study on temperature trends, snow cover and river discharge in the upper Gandaki shows that changing climate has an impact on snow

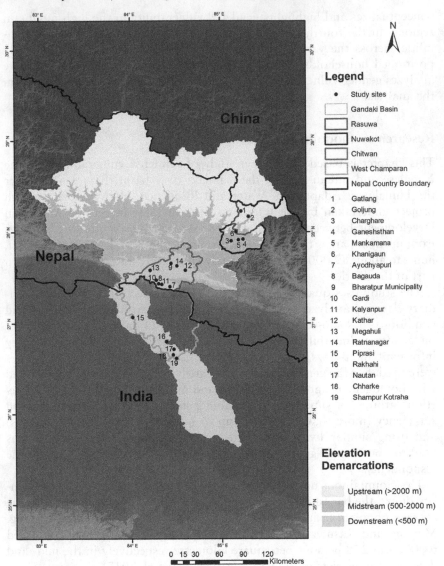

Figure 6.1 Gandaki river basin.

Source: Field study, 2017.

Note: This map does not reflect the political boundaries.

cover. Satellite data showing declining snow coverage between 2003 and 2012, correlated with temperature data between 2000 and 2007 and in-situ river discharge data between 1968 and 2010 indicate a trend towards increased river discharge (Gurung et al. 2017). The implications for farming are that soil moisture retained through annual snow coverage is declining, leading to crops drying prematurely. The farmers say that productivity of traditional crops is falling, leading to declining acreages of traditional high-value crops like buckwheat and millet. Between 2001 and 2011, the area under millet in high-altitude areas of Nepal like the upper Rasuwa has declined by 19 per cent, barley by 35 per cent and buckwheat by 40 per cent (CBS 2013). The impact of rising temperature is observed in day-to-day life, with women respondents reporting frequent incidences of nose bleeding among children during summer.

The three study villages in the upper Rasuwa – Gatlang, Goljung and Chilime[3] – are located along the Trishuli river, a tributary of the Gandaki about 2,000 m above mean sea level (Figure 6.2). People living in this rugged terrain eke out a living through snow moisture-fed high mountain agriculture and transhumance.

Tamang ethnic groups dominate the population. The second-largest is the Ghale ethnic group and there are a few marginalized Hindu castes of the *Dalit* group. According to the Village Development Committee data, Gatlang has 508 Tamang households and five *Dalit* households but no Ghale households. In Goljung, out of 269 households, 60 per cent are Tamang, 39 per cent are Ghale and there are only two *Dalit* households. In Chilime, out of 340 households, 20 are Ghales and five *Dalit*s. The rest are Tamangs. Tamangs and Ghales are homogeneous non-hierarchical Buddhist groups. They are Tibeto-Burman Mongolian groups and speak a similar tonal language. Historically disadvantaged and discriminated against, neither they nor the *Dalit*s are represented in government institutions or political parties

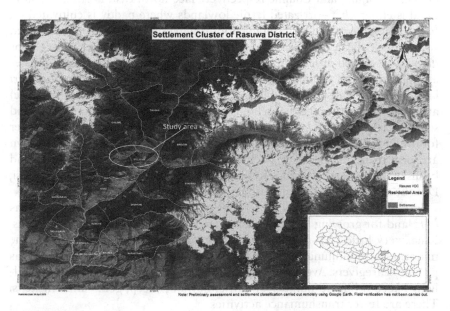

Figure 6.2 Satellite image of study villages in Rasuwa district.

Source: Created by ICIMOD, 2018.

at the national and provincial level (Tamang and Gurung 2014). According to the Civil Code of 1854, based primarily on Hindu religious and legal texts, the highest caste category is the holy thread wearer Brahmin/Chhetri, followed by the bulk of ethnic groups categorized as non-enslavable, such as Ghale, and enslavable, such as Tamang. At that time, the royal, non-beef-eating Ghales were considered as in a higher category than Tamangs. Over time these distinctions have become diluted with intermarriage between Ghale and Tamang (Campbell 1997). Below these is the 'untouchable' or *Dalit* category (Pradhan and Shrestha 2005). *Dalits* here are from the iron smith *(kami)* group. The presence of a few *kami* in each village was promoted by the state to support the farming community, here composed of Tamang and Ghale, by preparing farm tools in exchange for agricultural produce which ensured their food security. Hence, the *Dalits* have no farmland, and own only their homestead. A few Tamang and Ghale are large landowners, with holdings up to 0.6 ha., but the majority have on average 0.25 ha. land. Despite positive policies to address this historical social discrimination, *Dalits* are the most discriminated against, as they fall into the socially 'untouchable' category and lack land resources. Even the dominant Buddhist groups of Ghale and Tamang in the three villages follow historically rooted norms of untouchability.

The three villages differ in types of crop grown due to the differences in altitude, aspect and type of soil. The settlement in Gatlang is above 2,200 m above sea level, in Goljung 1,800 m and Chilime 1,400 m. The aspect of the villages also varies between windward and leeward side of the mountain.[4] Gatlang, Goljung and Chilime respectively face north-east, north and south. Land is either *khet* – irrigated fertile lowlands where paddy is cultivated – , *bari* – rain-fed terraced land suitable for growing cereals such as maize, millet, wheat, buckwheat, oat, linseed, potato, leguminous beans and vegetables – or *khar-bari* – thatch or fallow marginal land fit for trees, shrubs, thatch grasses and agro-forestry. Gatlang has very little *khet* and more *bari* or *khar-bari* land. There are substantial public grazing lands called *kharka* (range/pasture) at 4,500 m. On average, households in Gatlang have nine months of food self-sufficiency. Parvati Lake, located upstream, provides the village with water for drinking and for operating water mills for grinding. The main crops grown are local maize, black millet, buckwheat and local beans. Potato, barley and beans are grown as cash crops to sell in local markets at Chilime, Saybrubesi, Dunche and even the capital, Kathmandu.

In contrast, Goljung has more *khet* to produce seasonal paddy.[5] It has *bari* land for growing linseed, oats, potato, maize, brown millet and beans. It has very little *khar-bari* and no *kharka*. Access to drinking water is the major issue in Gojlung, with the burden falling on women in particular as primary caregivers. Average household food security is only six months. Villagers keep a limited number of cattle, buffalo, sheep, goats and chickens. There are fewer transhumance activities.

Chilime, at the bottom of the valley, has fertile *khet* land in the lower part and *bari* land in the upper part of the village, where millet, oats, potato and maize are grown. Lower Chilime is characterized by rapid urbanization. There are more hotels, lodges, business, trade and shopping facilities, as it has become the market centre for half a dozen new electricity plants along the Sanjen river. Schools, hospitals, banks and agri-collection centres as well as government offices are located here. Hence people have easy access to healthcare, agriculture, forestry, livestock and other relevant services.

Since farming alone is not enough, livestock also contributes to food security. Most of the grazing of cattle, yak and *chauris*[6] is transhumant. Among young people there is a growing trend towards alternative livelihoods due to the hardship associated with farming. Households with at least one family member (mostly men) working outside the village are 45 per cent, 80 per cent and 85 per cent in Gatlang, Goljung and Chilime respectively. Young people, especially men, also work on a seasonal basis as guides and porters in the nearby national park and the Tamang heritage trail.[7] Tourism is promoted and visitors, especially to Gatlang, have increased. Furthermore, the upper Rasuwa area has potential for hydro-electricity generation. In 2017, 16 hydropower projects were under construction in the district, generating employment opportunities for men and some literate women. It has also created economic opportunities: for example, 10 per cent of the equity in the 22 MW Chilime hydropower project is held by locals, including people in the three villages.[8]

Conceptualizing gendered vulnerabilities

Gender is considered here as a process by which various social attributes associated with being women and men shape the differential responses of groups to stressors, including climatic ones (Harding 1986). Socio-culturally constructed gender relations, roles and responsibilities determine differential access to, ownership of and control over resources for women and men across social groups and influence the ways they adapt to climate change (cf. Harding 1986; Goodrich et al. 2017). These resources may be social, human, economic, physical, financial or technological. We apply a sustainable livelihood framework to analyze various livelihood resources, taking heterogeneities and intersectionalities within a specific group into account while understanding gender vulnerabilities.

We do not confine our understanding of vulnerability to climate change to extremes like flood and landslides. Rather, we consider the impact of climate change on people's lives as a gradual onset of climatic stresses such as increase in temperature, decline in snowfall, erratic rainfall, and a gradual increase of the stresses caused by too much or too little water. Hence vulnerability is assessed in response to gradual processes of climatic stress, as well as climatic extremes and hazards like landslides, floods and droughts, including cataclysmic events such as the earthquakes of 26 April and 12 May 2015.

Understanding gender vulnerability is an opportunity to explore how people turn challenges into opportunities. People and communities are not just passive victims, but also active managers of vulnerability (IPCC 2014; Gilson 2013). Hence, integrated and multidimensional approaches are applied to identify its drivers and causes. Our study examines gendered vulnerabilities as the interplay of external factors with the existing internal environment. External factors such as market forces, increasing consumerism, urbanization, globalization, infrastructure development and technological interventions are contrasted with internal geo-political and socio-economic factors of social structure, gender, geography, political economy, decision-making processes and institutions (Goodrich et al. 2019b).

Gender roles and responsibilities

In this section we analyze gender disparities in roles and responsibilities that lead to differential gender vulnerabilities. The analysis is organized according to the major traditional and newly emerging livelihood strategies.

Transhumance herding, livestock and farming

Over the last three decades, transhumant herding complemented agriculture and was the over-arching livelihood strategy. However, the life of a herder is challenging.

A transhumant herder ascends to the pastures (*kharka*) from the settlement in April and descends by September every year. It takes him a fortnight to reach the meadows along with his septuagenarian father and 55 animals. His mother treks up every month to supply them with food, medicines, oil, salt, and fodder-grains for the animals. Once in the *kharka*, a herder wakes up at 3-4 a.m. and grazes the animals until evening. Then he counts and corrals the animals in the camp site and ensures their safety from wolves and snow leopards. Often the animals stray or go missing in sub-zero temperatures. He goes looking for them with his sheep dogs. Often, many are dead. Sometimes, he returns to his rudimentary shelter at midnight. Exhausted, he often eats raw oats with water to appease his hunger. A herder's life is not for the faint-hearted.

Due to the hardships involved and the dwindling supply of grasses, people have started to give up this work. Though the establishment in 1977 of a cheese factory in Gatlang that provided cash income was an incentive to continue herding, it was only recently that more farmers started to get involved in herding. Supplying milk to the cheese factory provides a more consistent income than the risks of farming. There are also government programmes to support herders, such as loans to buy animals.

Gatlang has about 110 Tamang households and Chilime has 80 households of which 60 Tamang and 20 Ghale families are herders. Goljung has just 10 households consisting of two Ghale and eight Tamang families. The Sanjen *kharka* grassland for herders from Chilime and Gatlang cannot accommodate all the *goths* (temporary caravan settlements comprising 50 animals on average). Water scarcity and competition for grazing lands are the major issues facing herders.

The herders are mostly middle-aged and older men, mostly those unable to migrate for labour work. Among the younger families, the men take the animals and stay in the *kharka*, while women journey up and down every month to supply food and look after children at home. In households where the men have migrated, both elderly parents go to the *kharka*. During our field visits, more elderly than younger women were seen in the *kharka*.

When the men are away in the *kharka*, the women bear the triple burden of the reproductive and productive work, as well as community management responsibilities. Women must regularly supply food and other necessities to the men in the *goth*. Households with ageing members are extremely vulnerable.

Negotiations between mother-in-law and daughter-in-law are challenging when women are to join the *kharka* for a longer period or to supply goods. A daughter-in-law explained, "I do not have good understanding of *kharka* like my mother-in-law. I take care of my small child. My mother-in-law supports our family members in *kharka*." The elderly women in the *kharka* complained, "The new generation is not interested in learning." These older women experience multiple hardships and health issues. One of them said, "I have extreme knee and back pain these days." In general, Nepalese women experience bone-related problems more than men due to lack of sufficient nutrition and rest during pregnancy and after childbirth (Lamichhane 2005). Milk consumption among these women drastically reduced once milk began to be supplied to the cheese factory. An educated young man who works as a tour guide and is actively involved in village institutions highlighted bone problems as an issue for elderly members of herder families, especially women. Previously, an internal barter system for milk and by-products like butter and local cheese provided sufficient nutrition.

Other than animals in transhumant herding, the women are primary caregivers. They are responsible for fodder collection, feeding animals, cleaning sheds, milking and providing traditional first aid to sick animals. The men do the buying and selling of livestock or products like meat, milk, leather/hides and bones. Women lack cash in hand compared to men. But involvement with small animals at home provides nutritious foods like milk, meat and eggs. *Dalit* women care for goats and chickens. *Dalit*s cannot afford to buy and rear big animals like cattle.

Timed task allocation analyses reveal that women in farm households perform most of the tasks related to agricultural and livestock management at the household level (Table 6.1).

Our data suggest that women are involved in more daily labour work than men. Another study suggests that in similar conditions, women out-perform men by nearly 3.5 hours a day on average (Shtrii Shakti 2015; Dough Merrey et al. 2018). Women go to bed late and get up early so they can get all their work done (Koirala et al. 2017).

Table 6.1 Broad task allocation study (TAS) in the case study sites

	Women	Men
Agriculture	Seed selection	Ploughing
	Field preparation	Hoeing
	Manure preparation	Harvesting
	Carrying manure	Barter trade
	Hoeing	Selling
	Plodding	
	Seeding	
	Planting	
	Weeding	
	Harvesting	
	Storing	
	Food processing	
	Grinding	
	Extracting oil	
Livestock	Collecting grass/fodder	Local herding
	Feeding animals	Transhumance herding
	Cleaning/maintaining sheds	Managing crossbreeding,
	Milking	reproduction
	Making butter/cheese	Culling and meat production
	Nursing young/sick animals	Buying and selling livestock
	Preparing/drying manure	Marketing milk, cheese, animal
		products

Source: Field data, HI-AWARE, ICIMOD, 2017.

For the *Dalit*s, being landless, the role chart is applicable to those who can lease unutilized low-fertility land from Tamang and Ghale households. Many of them are farm labourers. A few women have goats and chickens that provide cash or meat for the family. *Dalit*s who make and repair agricultural tools for Tamangs and Ghales receive 10 kg of cereals a year for each adult. However, cash payments for new tools at the local markets rate are catching on fast. Neither the kind nor cash contribution is enough to meet the *Dalit* household's annual food requirements. Some of these women shared that they had experienced domestic violence as a result of economic stress in the family.

In general, women are paid two-thirds of what men get for the same work due to prevalent notions of women being incapable of hard work. Most of the women labourers are *Dalit*s. In jobs like masonry, carpentry or construction, skilled men get four times what women are paid. This lowers the social and economic position of women compared to men. Reciprocal exchange of labour, the norm until 15 years ago, is gradually being replaced by cash payments due to market penetration in the villages, and this is gradually eroding the social ties that used to be a resource for facing hardship in the high mountains.

Eco-agro tourism, home stay and handicrafts

With traditional livelihoods no longer sufficing to meet annual household food requirements, many young men are exploring opportunities in tourism. The first is to the nearby Langtang valley mountain treks and Gosaikunda lake trails. The second is within the three study villages, where the Rural Poverty Alleviation Project (2000–2010) promoted the Tamang Heritage Trail, although the 2015 earthquake[9] halted momentum. The more educated take jobs as guides, whereas strong non-English-speaking men work as porters. As part of the project about 100 Tamang and 20 Ghale women were trained in ecotourism. Fifteen *Dalit* women received training in vegetable gardening and micro-enterprises. With the support of small grants, some started small enterprises, such as cultivating herbs as cash crops, woollen handicrafts, retail shops and homestays. Keeping a market-driven business going, however, requires the capacity to bear the financial risk and expand the business using social networks. It was mostly Ghale and a few Tamang women who could make use of these opportunities, as *Dalit* women have limited capacity for risk taking. It was mostly widows with social and financial capital and women financially supported by male migrant family members who could remain in business. These women entrepreneurs hired paid women labourers, mostly *Dalits*, to look after their livestock and undertake agricultural tasks.

Educated and wealthy Tamang and Ghale families benefited the most from the tourism industry. They invested in house building and renovation and used their contacts in the capital to attract tourists. Tamang and Ghale women also sell their handicrafts, while the men make furniture and houses for home stays.

Not all the villages offer equal opportunities for ecotourism. Of the three villages, Gatlang, which maintained Tamang cultural activities through a traditional institution called the *chogo* system that promotes social cohesiveness, attracted the most visitors. Gatlang has eight modest small-to-medium homestay lodges. In contrast, Goljung received the fewest visitors as trekkers often bypass it, going straight down to Chilime. A major tourist attraction in Chilime is the hydropower plant dam, which forms an artificial lake of over one square kilometre.

Under the *chogo* system, or the Council of Elders, five or six members are selected by community consensus to run village affairs relating to natural resources management, social events including births, deaths and marriages, and village development works. The head of the committee is known as *Chogo*. Traditionally, the scope of the committee included conflicts with neighbouring villages concerning wild animals. Only physically strong men were *Chogos* as the work required mobility as well as cooperative ability and the skills to handle confrontations. It is a voluntary position, and women can be members, but according to key informants, women rarely volunteer for the role due to their household responsibilities. During the study period, one woman member was observed to be overseeing social events. During festivals, villagers give *Chogo*s gifts such as scarves, shawls and cotton turbans in recognition of their services. Since 2015, in recognition of the importance of the role of *Chogo* in local social governance, the local government has allocated a monthly remuneration of NRs. 6,000 to each member, which has supported continuation of the *chogo* system in Gatlang. The twice-yearly week-long *mani rimdu* cultural festival, held during the birthday celebration of Lord Buddha in May and again during harvest festival in August, has given important cultural significance to this village, attracting tourists and creating alternative livelihood opportunities.

Migration, social capital, finance

Three decades ago, upper Rasuwa was a place isolated from national and global markets. The trek to the capital, Kathmandu, would take a week. This changed after the army built the 160 km Trisuli–Somdang jeepable 'dirt road' to access the Somdang lead-zinc mines. A highway linking Kathmandu to Rasuwagadi on the Nepal/Tibet border also increased mobility. With road connectivity, mobility increased, especially for young men who moved to cities like Trisuli and Kathmandu for work and study. Substantial out-migration began in 1990 when Nepal adopted neo-liberal policies. It increased during the Maoist-led civil war between 1995 and 2006. The then government introduced liberal passport policies to encourage young people to go abroad for work. Many young people from upper Rasuwa took jobs in the Gulf and South-east Asia to escape from political unrest.[10]Although it is mostly men who migrated, a few Tamang and Ghale women from Chilime and Goljung also migrated abroad, whereas women from Gatlang mostly went to the nearby villages at Trisuli and up to Kathmandu for work.

The impact of migration in upper Rasuwa is mixed. FGD participants mentioned that of the total number of families from which at least one member migrated abroad for alternative income, only half were able to uplift their socio-economic status, whereas the other half became more vulnerable after migration. The incidence of family breakdown and indebtedness to finance the emigration are high. The men who migrated abroad returned home after at least two years. This affected a Tamang woman from Goljung, who moved to a nearby marketplace at Syphrubesi to run a hotel. Leaders of a mothers' group stated, "After two decades of male migration, there are fewer men in the villages. The social costs of isolation and separation are high. There are incidences of promiscuity and family breakdown. Physical, social and psychological abuse and violence against women has increased. Sometimes, through our group we rehabilitate women from our villages. We often receive refugees such as pregnant women from southern villages too, who have escaped to the north due to shame. We provide them counselling, shelter and even bus tickets to return home." A young Tamang daughter-in-law mentioned, "We are left with managing our household chores, agriculture and livestock rearing. Social and political participation are also our duty now. We have no time at our disposal." Another young woman from Goljung whose husband is abroad said, "Gatlang had a huge forest fire five years ago that no one could put off. Landslides, soil degradation, drying of water system have made life hard." Among the migrant families, many men face challenges in new workplaces abroad, while women and older parents remain behind in a degrading environment without the support of young male family members.

None of the *Dalit* families have opted for foreign labour migration as they lack the financial resources required to pay upfront. Among the Tamang and Ghale, only those families with strong social and financial capital to absorb the initial financial and social cost were able to uplift their status. Families that have male relatives (such as brothers) to support women and the elderly in the day-to-day running of their livelihood activities could add resources after migration. Some of them invested the savings they had made after migration in farming medicinal herbs in Chilime. Others faced increased drudgery and indebtedness, paying huge interest on the loan taken out to send the family member abroad, and even family breakdown, rendering them still more vulnerable.

Trade and other economic activities

Emerging new development opportunities in the area such as hydropower and market penetration have provided the opportunity to trade in hydropower shares, medicinal herbs, small and medium food shops and local liquor. Involvement in such opportunities varies across groups and genders.

The Chilime hydropower project was the first in Nepal to provide 10 per cent of its shares to local communities as part of benefit sharing with local communities. Villagers from all three villages were eligible for these shares. A maximum of 300 shares could be allocated to one individual, at a share value of Nepali Rupees 100. But those who could not get to the hydropower office or were unavailable as they were in *kharka* could not buy the shares. The value of a Rs.100 share during the study period was about Rs. 1,650. The shareholders receive annual dividends of 15–20 per cent. Trading these shares is a new opportunity. Educated and politically active Tamang and Ghale men are actively engaged in share trading to build their assets. Many used their shares to repair homes damaged in the 2015 earthquake. Some bought motorbikes. Those who were less knowledgeable about shares sold them cheaply, losing a valuable asset. Among these are uneducated *Dalit* women and men, older people in the *kharka* and relatively poor Tamangs. At the household level, it is mostly men who control share dealings for both women and men. A mother of two boys aged 6 and 2 from Goljung relates how despite her higher secondary education, it is her husband from Gatlang who takes the major decisions on trading the hydropower shares. "My husband who travels frequently to Kathmandu carries out the share transactions for both us. I rely on him and accept his decisions. I hardly get involved in creating assets through my shares. Asset creation for Tamang, Ghale and *Dalit* women are from parental wedding gifts and jewellery bought from savings, and for the fortunate few through remittances and gifts sent by her spouse if he is a conscientious and loyal person."

Women are involved in the collection and processing of medicinal and aromatic herbs found in the forests. About 30 women in Gatlang, 55 women in Tetangche, Chilime and 50 women in Gonggang, Chilime from the Tamang community have formed cooperatives. They cultivate and process herbs as micro-enterprises. The trade in herbs supplements household income. A young woman said, "Forests are like our parents. We revere and worship them. We get resources to meet our social, economic and environmental needs. We worship forests as our mother on *Janai Purnima,* a full moon day in mid-August. The lake above the three villages is called the *Ama Chedingmo,* meaning 'our mother' in our language."[11] The forests surrounding the lake provide plants with medicinal value. Women used some of the plants for household consumption and others were bartered or sold to middlemen of ayurvedic companies. However, in summer 2012 the Gatlang forest was almost completely destroyed by a fire that raged for months.

The Tamang women of the three villages and the Ghale women of Goljung and Chilime bazaar run shops to sell groceries, medicines, stationery, beauty products and other household necessities. There are 12 such shops in Gatlang, five in Goljung and 30 in Chilime. The daily turnover of these shops ranges from NRs. 500 to NRs. 5,000. Due to its strategic location, the Chilime bazaar vendors enjoy the best business. The *Dalit*s are mostly consumers and have not been able to gain much from these new market opportunities. Both *Dalit* men and women are engaged in the labour market to ensure food security.

There are small and medium businesses that sell cement, sand, stones, steel rods, corrugated iron sheets, paint, wood, bricks and hardware accessories. Some Tamang and Ghale traders supply seeds and fertilizer to the villagers. They harvest and sell potatoes, leguminous beans and medicinal plants on a large scale. There are two such traders in Gatlang, one in Goljung and five in the Chilime bazaar. Their monthly turnover ranges from NRs. 100,000 to NRs. 1,000,000. The men are in charge of the business while women are looking after the family.

Every Tamang and Ghale household brews *raksi* a distilled liquor and *chaang,* beer from millet, maize, wheat and rice. All adult women, except the *Dalit*s, know how to brew liquor. Local alcohol is consumed daily in the household for hospitality and to meet spiritual needs. Women barter or sell it within the village to earn money to pay for children's education and to meet daily household needs. Sometimes savings from liquor sales are used by the women to buy jewellery and clothes. This economic activity is not only an essential part of life for the Tamang and Ghale women but an intrinsic part of the culture of the community.

Saving and credit activities

Facilitated by government and non-government agencies, women's saving and credit groups have been promoted in all three villages over the last 10 to 15 years. Previously a few wealthy landlords would lend the money needed for business, agriculture, livestock and daily subsistence at exorbitant interest rates, ranging from 60 to 100 per cent. The women in the saving and credit groups meet once a month to review turnover, cash flow and balance sheet. The amount loaned and the interest rate depends on the balance available and the type of business proposed by the women members. Each borrower has to pay 3 per cent interest a month, or 36 per cent annually, a higher rate than the local banks which charge 18 per cent. However, accessing loans from local banks is a complicated process for women as many of them have no collateral, so it is easier to approach the women's saving and credit group. Moreover, travelling to the bank is expensive for some. Whereas women entrepreneurs at Chilime with large investments can access bank loans, most of the middle-class women used loans from the women's group to buy seeds, fertilizer, agricultural inputs, livestock or poultry, or to construct simple greenhouses for vegetable and herb growing and drip-irrigation systems. Others used the loan to meet social needs, including children's school fees and stationery, medicines or emergencies like death or accidents. One of the women members added, "It is the women who come forward to borrow for the household requirements and often act as proxies to obtain loans for men's needs too." It is mostly middle class Tamangs and Ghales who benefit from these group savings systems. Many Dalit women fail to access such opportunities.

Gendered vulnerabilities, capabilities and adaptation

Among Tamang, Ghale and *Dalit* social groups in the study villages, those who lack physical, financial, human, natural and social capital are the most vulnerable. The *Dalit*s are the lowest ranking. After them come marginal Tamang and Ghale, including widows without brothers or male cousins. They tend to be the most vulnerable to climatic stressors. In Goljung, which lacks drinking water, women in this category bear the extra climatic stress burden of managing water for the family. Among the Tamang and Ghale, older people involved in transhumant herding are equally vulnerable to climatic stressors due to their age, health issues and the hardship involved in these activities. In terms of geography, women from poor household in Gatlang and the mountainous part of Chilime are more vulnerable to climatic stressors than those in Goljung considering that new irrigation will increase productivity and Chilime bazar is located in the valley bottom where there is industrial growth and emerging businesses.

The deeply ingrained psychology of 'untouchability' and class discrimination makes *Dalit*s vulnerable in this society even though the villages are dominated mainly by Buddhist Tamang and Ghale. Some poor Tamang and Ghale people, as well as *Dalits*, find day-to-day subsistence living hard to manage. As a coping strategy, many *Dalits* and marginal Tamang and Ghale have changed their religion to Christianity, as a number of evangelical Christian and faith-based Protestant churches have sprung up in all three villages in the past two decades. According to a key informant in Gatlang, "There are at least three churches in Gatlang and one each in Goljung and Chilime. I estimate that 10 per cent of the Tamang and Ghale population and almost all Dalits have embraced Christianity. This is due to more empathy, basic education and economic help forthcoming from the church groups to these households."

Among migrant Tamang and Ghale people, those who were able to increase their human and financial capital by working abroad can invest in new ventures like large-scale medicinal herb farming. There are few such families in Chilime, so the results are yet to be observed. It is possible that these large-scale enterprises will help to absorb the climatic stress on livelihoods by creating employment opportunities for other subsistence farmers. In contrast, those who lost both financial and social capital through working abroad have become more vulnerable, with women and men sharing an equal burden. Government intervention to create safety nets for families attempting to work abroad could avoid this increase in vulnerabilities. These measures would include monitoring agencies that supply human resources abroad, providing modest loans to support the processing cost; making both women and men aware of the pros and cons of foreign labour migration would help families to make informed decisions.

The process of building social capital through established community life and social mores among Tamang and Ghale, especially among the middle class, has helped to absorb the shock of various stressors, including climatic ones. Spiritual and cultural practices that promote kinship relations can provide immediate support. These practices include: ethics; a code of conduct and moral behaviour called *rimthim*; respect for the elderly; honouring guests (*salgar*); traditional self-help and barter system (*aicho paicho*); mutual labour exchange (*parma*); shamanism (*dhami/jhakri*) as faith healing and traditional medicine; festivals such as *Losar* (New Year) in January/February; festive celebrations and community carnival (*mani rimdu*) in May/August; and the institution of local Lama and Buddhist monasteries. In Gatlang, households' contributions for festivals are equitably shared, based on income. Maintaining the traditional practice of *chogo* in Gatlang while merging it with statutory institutions has created a harmonious continuation of traditional institutions while working with the state government. This has led to expansion of tourism in Gatlang and opportunities to diversify livelihoods. Well-informed women are taking advantage of these opportunities to become involved in handicraft micro-enterprises and running homestay lodges.

Conclusion

This chapter has shown that there are gender differences in role allocation, access to and control over resources, and dependency relations in Tamang, Ghale and *Dalit* communities. Women contribute significantly as farm labourers, whereas financial work is done by men. Women are paid two-thirds of men's wages. For work outside the home, women are dependent on men. Living in a fragile environment, social capital has a high value and women are highly dependent on men for it. The reciprocal effect of this relationship is that men's access to and control over most of the livelihood resources and decision making is high. Women's dependence on men and limited mobility, and the low value placed on women's work makes women more vulnerable while responding to stressors than men.

The level of vulnerability and resilience among women differs among different ethnic and caste groups depending on class, age, education level and geographical location. Women from the *Dalit* caste group, who are the lowest in the social hierarchy and have limited livelihood resources, bear multiple burdens of climatic stressors. Among Tamang and Ghale, elderly women are more vulnerable while supporting transhumance herding in harsh environments. The vulnerability of women in Tamang and Ghale households where men have migrated is high among economically poor households with low education levels and limited social networks.

The emerging opportunities in tourism were also found in the study to be differently used by women due to their differing abilities to bear risk and opportunities. Educated and economically stable Tamang and Ghale women are involved in enterprises that have helped to absorb the climatic shocks affecting farming-dependent livelihoods. Women in Goljung managing water for the family bear additional climatic stress burdens, while urbanization has provided some women in Chilime with more entrepreneurship opportunities than in the other two villages.

Policies and programmes that offer adaptation solutions to address vulnerabilities need to recognize the fact that within caste, ethnic and gender groups there are different ways of responding to stressors depending on individuals' levels of access to and control over resources, and their associated capabilities. Initiatives aimed at addressing climate vulnerabilities need to address vulnerabilities caused by intersectional issues of social, financial, economic, political and physical resources. These initiatives may be: interventions to support access to drinking water, irrigation facilities or rural technology for agricultural and livestock rearing; sustainable soil fertility; forest and environmental protection; conservation of biological diversity; improved agriculture and livestock breed management; improvement of rangelands and improved animal breeds; introduction of horticulture for high-value fruits, nuts and medicinal plants; or non-farm sectors like eco-tourism, hydropower share trading, migration and employment generation.

Notes

1 Nepal was previously divided into five development regions, 14 geographical zones and 75 districts spread over three ecological regions of mountains, hills and plains called the Tarai. The districts had more than 4,000 villages with the Village Development Committee (VDC) as the lowest political and administrative unit. This demarcation was changed in early 2018 in the spirit of the Constitution of Nepal, 2015. Following this change, the country consists of five development regions, 14 zones, seven provinces from east to west with 713 rural and urban municipalities including five metropolitan cities, and 77 districts. (*Nepal Gazette* 2018).

2 See https://hi-aware.org/

3 Under the Federal Republic of Nepal's latest political configurations, these three villages have been subsumed and amalgamated into one rural municipality called Ama Chidingmo.

4 In meteorology, leeward and windward are technical terms for the directional sides of a mountain. The windward side is the side which faces the prevailing wind (upwind), whereas the leeward or lee side is the side sheltered from the wind by the elevation of the mountain itself (downwind) (Tiffany, 2018).

5 There was a government-supported irrigation facility that failed to function after construction. Following repairs it started to function in 2017 but the outcome of the system is yet to be observed.

6 A cross between a male yak *(Bosgrunniens)* and a cow *(Bosindicus)* is called a *chauri*. *Chauri* is a hybrid female which produces more milk and is preferred. The male from this cross is called a *jopo (djo)* and the *chauri* is called a *djimo*. The *jopo* is not fertile. There are many variations of hybridization of yak and cattle, but the most popular is the *chauri* among the transhumance herders of Rasuwa.

7 See https://thehimalayantimes.com/business/tamang-heritage-trail-reopened/

8 See http://www.chilime.com.np/images/supportive_docs/Annual-Report-10-11.pdf

9 See http://seismonepal.gov.np/strong-motion

10 The mid-2019 figures for labour migration from Nepal stood at 8 million young people, 90 per cent of them men. Source: South Asian Regional Trade Union Council, Kathmandu

11 The term *Adivasi Janajati* or 'original settlers' is a term given to over 100 Tibeto-Burman caste groups to distinguish them from the Indo-Aryan Brahmin/Chhetri and *Dalit* groups. The former group constitutes 58 per cent and the latter three around 42 per cent of Nepal's 30 million people. The Tamang and Ghale belong to the *Adivasi Janajati* group (*Nepal Gazette* 2018).

References

Campbell, B. 2017. 'Encountering climate change: Dialogues of human and non-human relationships within tamang moral ecology and climate policy discourses', *European Bulletin of Himalayan Research*, 49: 59–87.

Campbell, B. 1997. 'The heavy loads of Tamang identity.' In David N. Gellner, Joanna Pfaff-Czarnecka and John Welpton (eds) *Nationalism and Ethnicity in a Hindu Kingdom. The Politics of Culture in Contemporary Nepal*. Abingdon: Routledge: 205–236.

CBS. 2013. *Summary of National Agriculture Census 2011 AD (2068BS), 2013.* Kathmandu: Central Bureau of Statistics.

Dandekhya, S., England, M., Ghate, R., Goodrich, C.G., Nepal, S., Prakash, A., Shrestha, A., Singh, S., Shrestha, M.S., Udas, P.B. 2017. *'The Gandaki Basin: Maintaining Livelihoods in the Face of Landslides, Floods, and Drought.'* HI-AWARE working paper 9. Kathmandu: ICIMOD.

Gerlitz, J.Y., Hoermann, B., and Hunzai, K. 2011. *'Understanding Mountain Poverty in the Hindu Kush-Himalayas: Regional Report for Afghanistan, Bangladesh, Bhutan, China, India, Myanmar, Nepal, and Pakistan.'* Case Study. Kathmandu: IFAD.

Gilson, E. 2013. *The Ethics of Vulnerability: A Feminist Analysis of Social Life and Practice.* New York: Routledge.

Goodrich, C.G., Mehta, M., and Bist, S. 2017. *'Status of Gender, Vulnerabilities and Adaptation to Climate Change in Hindu Kush Himalaya: Impacts and Implications for Livelihoods and Sustainable Mountain Development.'* ICIMOD Working Paper 2017/3, International Centre for Integrated Mountain Development, Kathmandu.

Goodrich, C.G., Prakash, A., and Udas, P.B. 2019a. 'Gendered vulnerability and adaptation in Hindu-Kush Himalayas: Research insights', *Environmental Development*, 31: 1–8.

Goodrich, C.G., Udas, P.B., and Larrington-Spencer, H. 2019b. 'Conceptualizing gendered vulnerability to climate change in the Hindu Kush Himalaya: Contextual conditions and drivers of change', *Environmental Development*, 31: 9–18.

Gurung, D.R., Maharjan, S.B., Shrestha, A.B., Shrestha, M.S., Bajracharya, S.R., and Murthy, M.S.R. 2017. 'Climate and topographic controls on snow cover dynamics in the Hindu Kush Himalaya', *International Journal of Climatology*, 37: 3873–3882.

Harding, S. 1986. *The Science Question in Feminism.* Ithaca, USA and London, UK: Cornell University Press.

HI-AWARE. 2017. 'Socio-economic Assessment Report – Gandaki River Basin, Nepal.' Himalayan Water and Resilience Research (HI-AWARE) International Centre for Integrated Mountain Development (ICIMOD), unpublished report.

Humagain, K., and Shrestha, K.K. 2009. 'Medicinal plants in Rasuwa district, central Nepal: trade and livelihood', *Botanica Orientalis: Journal of Plant Science*, 6: 39–46.

IPCC. 2014. *'Climate Change 2014: Synthesis Report. Contribution of Working Groups I, II and III to the Fifth Assessment Report of the Intergovernmental Panel on Climate Change'* (Core Writing Team, R.K. Pachauri and L.A. Meyer (eds). Geneva: IPCC: 151.

Khan, A., Cundill Kemp, G., Currie-Alder, B., and Leone, M. 2018. 'Responding to Uneven Vulnerabilities: A Synthesis of Emerging Insights from Climate Change Hotspots'. CARIAA Working Paper no. 22. International Development Research Centre, Ottawa, Canada and UK Aid, London, United Kingdom. Available online at: https://idl-bnc-idrc.dspacedirect.org/handle/10625/56958

Koirala, M., Udas, P.B., and Goodrich, C.G. 2017. 'Gender and social inclusion in environmental discourse'. In D. R. Bhuju, M. Koirala and R. Nakarmi (eds), *Environmental Science- Some Theoretical Background and Applications.* Kathmandu: Central Department of Environmental Science, Tribhuvan University and International Centre for Integrated Mountain Development.

Lamichhane, A.P. 2005. 'Osteoporosis-an update'. *Journal of the Nepal Medical Association*, 44(158): 60–66.

Merrey, D.J., Hussain, A., Tamang, D.D., Thapa, B., and Prakash, A. 2018. 'Evolving high altitude livelihoods and climate change: A study from Rasuwa District, Nepal', *Food Security*, 10(4):1055–1071.

Nepal Gazette. 2018. Department of Printing, Ministry of Information and Communication, Government of Nepal

Pasteur, K. 2011. *From Vulnerability to Resilience: A Framework for Analysis and Action to Build Community Resilience*. Rugby: Practical Action Publishing.

Pradhan, R., and Shrestha, A. 2005. *Ethnic and Caste Diversity: Implications for Development*. Kathmandu: Asian Development Bank, Nepal Resident Mission.

Sharma, E., Molden, D., Rahman, A., Khatiwada, Y.R., Zhang, L., Singh, S.P., Yao, T., Wester, P. 2019. 'Introduction to the Hindu Kush Himalaya Assessment.' In P. Wester, A. Mishra, A. Mukherji, A.B. Shrestha. (eds) *The Hindu Kush Himalaya Assessment: Mountains, Climate Change, Sustainability and People*. Berlin and Cham: Springer.

Shtrii Shakti. 2015. *Revisiting the Status of Women in Nepal*. Kathmandu: Shtrii Shakti.

Tamang, M.S., and Gurung, O. 2014. *Social Inclusion Atlas of Nepal, Ethnic and Caste Groups* (Vol. 1). Central Department of Sociology/Anthropology, Tribhuvan University, Kathmandu, Nepal.

Part II
Adaptation and Wellbeing

7 Wells and well-being in South India[1]

Gender dimensions of groundwater dependence

Divya Susan Solomon and Nitya Rao

Introduction

In 2010, nearly 27 million hectares of land in India were groundwater irrigated compared to 21 million hectares irrigated by surface-water sources (Mukherji et al. 2013). Groundwater usage is growing at an unprecedented rate; it is estimated that one in every four rural households owns at least one groundwater irrigation structure (Shah 2013).

Groundwater development has been credited with increasing farm incomes and well-being by increasing productivity (Roy and Shah 2002; Sekhri 2014). On the other hand, it has also resulted in chronic resource depletion in terms of both quality and quantity (Gleeson et al. 2010). Groundwater irrigation is currently at an impasse. It has cemented its vital position in the drought-proofing of agriculture. However, the exploitation of the resource for irrigation has resulted in critical groundwater levels, particularly in already water-stressed regions (Kumar and Singh 2008; Livingston 2009). The number of irrigation blocks in India reporting overexploited groundwater levels grew at an alarming rate of 5.5 per cent per year between 2002 and 2007 (Gandhi and Namboodiri 2009). Increased well proliferation has led to aquifer contamination and salinization, alongside increased pumping costs. The problem is particularly severe in arid and semi-arid regions, where communities depend on groundwater for both domestic and agricultural purposes (Taylor 2013).

Water scarcity is expected to grow in the future, exacerbated by changes in precipitation patterns and increase in temperatures (Shah 2013). A study by Geethalakshmi et al. (2011) on the impacts of climate change on the Cauvery basin in Tamil Nadu, using regional climate models, showed an increasing trend in maximum and minimum temperatures, and a decrease in the number of rainy days. These climatic changes will have an impact

1 This work was carried out under the Collaborative Adaptation Research Initiative in Africa and Asia (CARIAA), with financial support from the UK Government's Department for International Development (DfID) and the International Development Research Centre (IDRC), Canada. The views expressed in this chapter are those of the authors and do not necessarily represent those of DfID and IDRC or its Board of Governors. Our sincere thanks to the internal reviewers, interview respondents and field staff.

on the hydrological cycles in the region and lead to more run-off and less recharge, affecting the groundwater tables. There has also been an increase in the frequency of droughts in the state. Tamil Nadu declared droughts both in 2016 and 2017 (Promod 2017). This has driven farmers to increase their dependency on groundwater to secure their crops. Climate changes act as a force multiplier, increasing the criticality of the resource while simultaneously threatening it (Shah 2013).

The emergence and spread of an intensive groundwater-based irrigation regime has had important socio-economic consequences for rural households (Roy and Shah 2002; Shankar et al. 2011). Curiously, given the centrality of groundwater to the contemporary agricultural economy, few studies have focused on the socio-economic impacts, particularly gendered intra-household impacts of groundwater usage and dependence (Mukherji and Shah 2005). In fact, a recent review of the literature on water and gender in India (Kulkarni 2016) focused entirely on surface water and the gendered and social facets of its management. In our research, we attempt to fill this gap by focusing on the relationship between, and the micro-politics of, gender and the multiple uses and users of groundwater, in both agriculture and the domestic sphere. Specifically, we focus on the everyday experiences of men and women to understand the gendered segmentation of agricultural labour and its relationship to the control of assets, including water.

Context and methodology

With only 3 per cent of the national water resources, Tamil Nadu is one of the most water-vulnerable states in India (Janakarajan 1999). Around 15 per cent of the land area of Tamil Nadu is in the semi-arid zone, with less than 700 mm annual rainfall (Walker 2012). This explains why 56.8 per cent of the net sown area in Tamil Nadu was irrigated in 2009–10 (DES 2011).

Dependence on surface water (tanks and canals) for irrigation has been replaced by extensive groundwater use in the state (Kajisa et al. 2007); the area irrigated by groundwater has doubled since 1950 (Janakarajan and Moench 2006). Fifty-five per cent of the irrigated land is under groundwater in Tamil Nadu (DES 2011). Since 1991–92, state government subsidies to promote irrigation – particularly free electricity for pumpsets and loans for deepening existing wells and constructing new ones – have reoriented the role of wells from a source of supplemental irrigation to a key productive asset (Kannan 2013).

Lack of appropriate and properly enforced state regulation promotes fierce competition among farmers to extract groundwater to cultivate water-intensive cash crops (Phansalkar and Kher 2006). From 2002 to 2012, the groundwater declined by 1.4m annually on average, indicating that groundwater usage was nearly 8 per cent more than the annual recharge rate (Chinnasamy and Agoramoorthy 2015). Richer farmers can drill deeper, giving them an advantage over small and marginal farmers, who are burdened with 73 per cent of the non-functioning wells (Kulkarni et al. 2015). Furthermore, extensive ground and surface-water pollution by textile and paper mills has seriously affected the the availability and quality of drinking water, with adverse consequences for human health and well-being (Puthiyasekar et al. 2010; Saravanakumar and Kumar 2011).

This chapter is based on 25 in-depth interviews with 20 women and five men and a 200-household survey containing gender-disaggregated sections on livelihoods, time use and asset ownership, conducted in 2016 in two villages in the Coimbatore district of Tamil Nadu (Figure 7.1). Coimbatore, a drought-prone semi-arid district, was known for the cultivation of dryland crops over large areas using tank irrigation in the 1950s, but since then, along with the production of cotton and rice, the region has developed into a major textile hub (Kosalram 1973). Between 2003 and 2014, the cultivated area in Coimbatore declined, largely attributable to the scarcity of both water and labour (Selvaraj and Ramasamy 2006; Prabu and Ahmad Dar 2018a).

For the household survey, we selected households that stated agriculture as their primary livelihood strategy and owned agricultural land. Stratified proportional sampling based on landholding size (small: 0–2 acres;

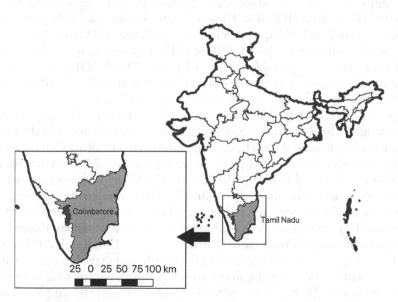

Figure 7.1 Location of Coimbatore district in India.

Data source: Global Administrative Areas (2012). GADM database of Global Administrative Areas, version 2.0. [online] URL: www.gadm.org. Map prepared by author.

medium: 3–6 acres; large: 7 acres and above) and caste, to ensure weighted representation, was used. Two panchayat[1] villages, Marudur (50 per cent small, 43 per cent medium, 7 per cent large) and Chinnakallipatti (47 per cent small, 45 per cent medium, 8 per cent large), were selected for the surveys (see Table 7.1). Although the surveys were primarily conducted with the head of the household, gender-specific questions were addressed to other members of the household. Only 8 per cent of the primary respondents were women, reflecting the lack of female-headed farming households in the region.

Twenty-five households were selected for in-depth interviews based on landholding size to understand the intra-household dynamics of groundwater usage among different economic classes of farming households. While the surveys provided information on the broader dimensions of livelihoods, in-depth interviews were used to understand groundwater usage, dependency and intra-household relations. Additionally, two focus group discussions (FGDs) with men only and two FGDs with women only were conducted in Chinakallipetti and Marudur respectively. Each FGD had 7–14 participants, though the caste composition was mixed. A majority of participants belonged to the Other Backward Castes (OBC). The names of interviewees have been changed to protect identities.

Chinnakallipatti and Marudur are small agricultural villages in Coimbatore district, having settlements of about 1,000 households. The majority of landowners are VellalaGounders, Mudaliars, VokkaligaGounders and VettaiyarGounders, all OBCs. These villages also have a small population of largely landless Pallars, who belong to the Scheduled Castes (SCs),[2] and Kurumbas, who are a Scheduled Tribe (ST), with small and medium landholdings. After the land distribution of the 1970s, the OBC castes established their hegemony in the region; they are the main landholding castes and occupy the majority of seats in the gram panchayat.[1]

Over the last decade, there has been a shift to intensified cash crop cultivation, aided by the groundwater revolution, mechanization and the introduction of hybrid seed varieties. Eighty-four per cent of farmers reported using groundwater for irrigation in the survey; of these 74 per cent reported at least one borewell failure in the last decade. While buffering against climate risks, such water-intensive cultivation has simultaneously increased the vulnerability of both the resource and the communities dependent on it. Secondary sources of livelihood involve livestock, small ruminants and agricultural labour (Coimbatore District Statistical Handbook 2012–2013, 2013–2014). As the region is close to the textile and manufacturing centres in Coimbatore and Tiruppur, many young people from the village migrate daily or seasonally for work. Several factories specifically employ young unmarried women (Heyer 2016) and pay them a lump sum for their dowry after three years' service in addition to their monthly wage (Rao 2014). Although the controversial Sumangali scheme is beyond the scope of this

Table 7.1 Land and well ownership among households, sampled by caste

Caste	Subcaste	Sample size	Average land holding size (acres)	Percentage with irrigation	Number of borewells (average)	Percentage with outstanding debts	Percentage that practice seasonal migration
OBC	VokkaligarGounder	48	5.2	90.8	3.2	70.8	70.8
OBC	Mudaliar	22	5.7	92	3.8	72.7	68.2
OBC	Chettiyar	10	6.3	97	4	60.0	60.0
OBC	VellaraGounder	45	3.6	89	4.2	84.4	80.0
SC	Pallar	37	2.4	84	2.8	89.2	81.1
ST	Kurumba	38	2.2	72	1.5	92.1	71.1
Total sample		200					

Source: Primary survey data collected in 2016.

chapter, the rise in dowries can be linked to the need for capital to dig additional and deeper borewells to sustain commercial agriculture.

Women play an essential role in the agrarian system here. However, the labour participation and responsibilities of women in the field and at home depend on the caste of the household. Rao (2014) and Heyer (2016) show that *Dalit* women are withdrawing from paid employment and moving to unpaid household work, often on farms, at least temporarily when their children are young, while OBC women are pursuing higher education, and at the same time contributing to household agriculture. Women also contribute to household nutrition by maintaining sustenance crops such as millet, vegetables and fodder either through intercropping or in domestic gardens. Despite women's role and contribution to farming, they remain subordinate to their male kin with respect to agricultural decision making.

Our household survey indicated that men, particularly from OBC groups (97.8 per cent), are the primary decision makers with respect to cash crops, including crop choices, irrigation inputs and markets. While still high, male control over decision making was relatively lower in SC households (94.2 per cent) and lowest in ST households (90 per cent). Men also carried out most technology-related and, indeed, more remunerative tasks. In the next sections, we briefly set out the changes in cropping patterns as a result of changing climate, policy signals and people's aspirations, and discuss the implications for livelihood choices. We particularly focus on the intersections of gender, caste and class which mediate access to and control over assets (land, livestock and money) and how this shapes intra-household cooperation and conflicts, considering also the special case of female-headed households. In the concluding section, we explore what the growing dependence on groundwater means for gendered well-being in general, and for specific caste and class groups.

Cropping patterns and divisions of labour

Tamil Nadu gets two monsoons, unlike many other parts of the country. The north-east monsoon, which arrives in early November, is the primary monsoon in the region and marks the peak *aadi* cultivation season. The south-west monsoon, which extends from the end of May to August, is comparatively weaker and marks the *avani* cropping season. A 2014 study found a significant shift in monsoonal patterns, with a weakening north-east monsoon and a strengthening south-west monsoon (Krishnaswamy et al. 2015). In our survey, 98 per cent of the respondents felt that total annual rainfall has decreased from 2005 onwards, and 84 per cent reported that rainfall variability had increased over this period. Irregular precipitation, along with increasing instances of drought (Kabirdoss 2017), has prompted farmers to adapt by shifting from supplemental to complete groundwater irrigation. A functional borewell hedges the household's fortunes against the capricious nature of monsoons and provides a reliable source of water.

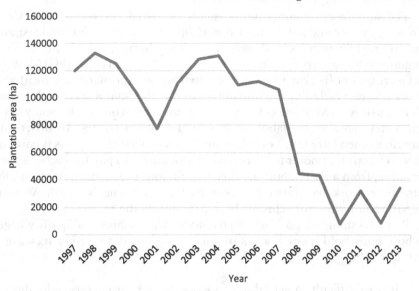

Figure 7.2 Plantation area of dryland crops (coarse millets and grains) in Coimbatore District, 1997–2015.

Data source: Directorate of Economics and Statistics (DES), (1997-2015) Government of India.

The expansion of groundwater irrigation has, in turn, led to a dramatic shift in cropping systems and crop management practices. It has facilitated the shift from rain-fed food and fodder crops (sorghum, pearl millet, foxtail millet and finger millet), to high-value but high-risk perennial cash crops like banana, turmeric, jasmine, tobacco and, recently, curry leaf (Figure 7.2). There was a decrease of 71 per cent in the cultivated area of rain-fed food and fodder crops in the district between 2000 and 2013 (DES 2000, 2013). Rain-fed millets are hardy and resistant to long dry spells, do not require expensive agricultural inputs and provide nutritional security to households. However, the market for these crops is limited and returns are low.

Around 72 per cent of respondents reported that they grow only perennial cash crops in 50 per cent of the cropped area. This shift has increased the risks associated with farming. Cash crops are input intensive and require considerable financial outlays. Almost 78 per cent of total respondents reported significant losses from agriculture in the last five years, mainly due to drought and water scarcity. This was fairly consistent across caste groups – OBC 74.6 per cent, ST 76.2 per cent, SC 79 per cent – although the losses were significantly larger for SC and ST farmers (74 and 73 per cent of investments on average) than for OBC farmers (61 per cent). Being largely small and marginal landholders, the SC and ST farmers depend on loans for inputs. Sources of loans include banks (24 per cent), moneylenders (36 per cent), and relatives (40 per cent), with interest rates of 4–7 per cent, 20–40 per cent and 10–30 per cent, respectively. The implications of this are discussed later in the chapter.

The shift in cropping patterns entails gendered costs and risks. Cash crops are water and labour intensive. Crops such as curry leaf and jasmine require regular pruning and weeding, and are harvested by hand. Labour requirements have tripled since the introduction of cash crops (FGDs). With an increase in off-farm employment opportunities (primarily for men) and work schemes including the Mahatma Gandhi National Rural Employment Guarantee Act (MGNREGA),[3] primarily used by women, the study area has witnessed a severe labour deficit for farming activities. To offset this deficit, women have taken on additional tasks in agriculture. This is particularly relevant for poorer farmers who cannot afford to pay the higher wages resulting from a tight labour market. Continuous engagement in physically strenuous work has, however, taken its toll on women's health. Women reported suffering from chronic back pain during the FGDs.

Rengalakshmi (41), a VokkaligaGounder from Chinnakallipatti village, whose household owns 6 acres of land, complained about her increasing work burden:

> It is so difficult to get labour to work on our farms, especially during the harvest season. When they do come, they demand too much money for us to pay … I have started working more on the farm now. I do not have a choice. My husband goes for construction work in the neighbouring town, and if I do not do agricultural work the crops will die.

Well-off farmers with larger landholdings use technologies such as drip irrigation to reduce labour requirements; but medium and small farmers depend on women in the household for their additional on-farm labour requirements. Women across castes play an active role in irrigation activities in the region. Around 47 per cent of total women in our study reported that they handled the majority of irrigation activities on the farm. For households with medium-sized holdings, mainly OBCs, 82 per cent of women reported that the average time available for household activities has reduced from eight hours to six hours per day over the last ten years. While borewells have ensured the availability of household water, reducing domestic drudgery for women, the time saved is often reallocated to unpaid farm work.

Manual methods such as flood irrigation cause increasing evaporation, and consequently a demand for more water and more labour. Alongside the increase in the on-farm unpaid labour of women, studies have also shown a positive impact on gender wage differentials (Narayanamoorthy and Deshpande 2003): with an increase in demand for labour, women have been able to negotiate higher daily wages for their agricultural work. This has been particularly significant for ST and SC women with marginal and small landholdings, 75 per cent of whom reported engaging in agricultural labour outside their farm. Lakshmi (35), a Kurumba female agricultural labourer from Marudur, explained:

Even 5 years ago, we were paid much less than men. Women would get about Rs. 100 per day for work on the field while men would get nearly Rs. 200. Now, since there is so much work on the fields and so few people, we can demand more wages ... Men get Rs. 350 and women get Rs. 300 for one day's work.

Around 78 per cent of respondents reported that they no longer grow any food crops. The declining production of nutritious millets and vegetables at farm level has led to increasing reliance on markets and the Public Distribution System (PDS).[4] The PDS in Tamil Nadu is among the most efficient in the country, with close to universal coverage and a relatively effective delivery system (Anuradha 2018). The security that the PDS provides is one of the reasons that households can make the shift away from food crops. However, the PDS has further entrenched the dietary shift away from millets and coarse cereals to low-nutrition polished white rice as the most consumed cereal crop.

Reflecting on the dietary changes in her household over the 10 years (2006–16), Revathy (53), a VellalaGounder from Marudur village with a 4-acre farm, said:

Our profits from agriculture have increased, but now we do not grow *ragi* (finger millets) or *jowar* (sorghum). The rice we eat does not give us the energy to work in the fields all day. Earlier we used to eat *ragimudde* (ragi ball) and not have any health problems; now almost everyone in the village has diabetes and other problems ... I ask my husband to buy *ragi* when he goes to the market, but he says it is expensive. Moreover, it takes a lot of time to prepare and my children prefer rice.

Revathy indicated that the reduction in cultivation of coarse cereals on the farm has reduced her autonomy in making choices on household nutrition. Nutritional choices are now based on market prices and availability.

Gendered assets

The bargaining power and agency of men and women in the household are shaped by their control over material assets and perceived (economic) contributions to the household (Sen 1990; Kelkar and Jha 2016). Access to and control over assets, however, are not necessarily determined by legal ownership, but are shaped by larger socio-cultural contexts and prevalent notions of legitimacy (Rao 2017). In Tamil Nadu, state-led gender discourses reflected in state-sponsored welfare schemes targeting women have strongly influenced notions of gendered asset control and ownership. We discuss here the impacts of borewells on two highly gendered assets: gold and livestock.

Gold

The total cost of digging a borewell in Tamil Nadu averages between Rs. 100,000 and Rs. 150,000, a substantial amount in a region where the average monthly income of families is around Rs. 5,000–10,000.[5] Around 62 per cent of the farmers reported having sold household assets to finance well boring; of them, 72 per cent had medium and small landholdings and 68 per cent belonged to ST. Kurumba households, which generally have lower incomes, are forced to sell household assets for large investments including in borewells.

Around 85 per cent of all households said that gold was the primary asset to be pawned or sold. Gold ornaments are a strongly gendered asset in the region (Swaminathan et al. 2012), traditionally given as dowry, a gift to a daughter on marriage. While dowry has been used to control and exploit women, women in in this region (western Tamil Nadu) have control over their gold ornaments. Gold has social legitimacy as a woman's asset. As well as a source of security that can be pawned in emergencies, it enhances her bargaining power within the household.

Jaya (48), a Pallar woman whose household owns 4 acres of land in Marudur village, points to her wedding photo and describes the dowry she brought:

> My parents gave 50 *pauns*[6] of gold for my wedding. It was not demanded by my in-laws, but my parents chose to present it to me. I have kept it safely for the last twenty years, but over time I have had to pawn most of it. Two years ago (2014), to dig a new borewell, my husband pawned nearly all of it. I was very upset; I wanted my daughter to have it for her wedding … Now I have only this *thali* (wedding pendant).

Jaya resented her husband pawning her gold, hinting that she was made to part with it forcibly. She comforts herself, however, with the reflection that this is a common occurrence in the village. With borewells increasingly seen as essential for a secure livelihood, women's gold is now being used to finance this investment. Jaya still worries about gold for her daughter's marriage and has started a small chit fund (a rotating savings system which yields a lump sum payment) from her agricultural labour earnings to buy gold.

Livestock

Livestock rearing is an important source of supplemental income for agricultural households, and provides nutritional security. The survey indicates that livestock rearing can provide as much as Rs. 3,000 per month, a buffer against crop failure and periods of declining agricultural productivity. Livestock ownership is strongly gendered owing to cultural norms and various government subsidies to promote the dairy industry. The ruling political party, All India Anna Dravida Munnetra Kazhagam (AIADMK), launched a scheme for the free distribution of milch cows and goats/sheep to poor women in 2011. At the time of the survey in 2016–17, about 36,000 milch cows had been distributed in the district under this scheme (GoTN, n.d.). Around 84 per cent of women across all castes in the survey reported that up to 60 per cent of their time each day is devoted to livestock-rearing activities, with women over 50 spending significantly more (up to 80 per cent of their time). Unable to undertake rigorous farm activities, they contribute to the household by looking after livestock. It is more or less a female domain.

Findings from our survey and FGDs reflect insights from studies on gender and livestock in India which highlight the role of livestock and small ruminants in the managerial and economic autonomy of women (Upadhyay 2005). Women handle the processing of milk into ghee, butter and other dairy products, and maintain personal savings through the sale of these products. The prevalence of women's milk cooperatives in the region supports and reinforces the importance of livestock as a source of women's income, to be spent on their personal or children's needs. Mariamma (62) is a VellaraGounder woman who has lived with her brother's family since her husband died when she was 20. She said: "I received a cow as part of a government scheme in 2008. I take complete care of the cow including taking it to graze and milking it. In this way I can contribute to the family." The area still has some common pasture land in the surrounding forested areas. Meanwhile the larger incomes from crop sales are spent on consolidating agricultural investments, seen as the responsibility of men as 'providers' (Garikipati 2009).

Rearing livestock is a water-intensive activity, as dairy cows consume up to 171 litres of water per day (Schlink et al. 2010). With lakes and ponds in the region shrinking, most of the water for livestock is provisioned from borewells. During periods of water scarcity, there is fierce competition within the household over limited water resources. Men prioritize water for maintenance of cash crops, while women prioritize the maintenance of cattle. Women reported that at times of extreme scarcity, such as in 2013 when the north-east monsoon failed, they were unable to maintain their cattle and had to resort to distress sales. This was higher among SC women (82 per cent) who could not afford to buy water, unlike OBC women who could buy water from neighbours' wells. ST women resorted to using water sources in forested areas, and although this further impinged on their productive time, they managed to maintain their livestock.

Despite the importance of livestock for women's livelihoods, men view livestock as a disposable asset. This was highlighted in our survey, with 82 per cent of male household heads reporting that cattle are sold in case of financial stress. Further, the area devoted to fodder crop production reduced by 66.87 per cent between 2000 and 2013 (DES, GOI). The reduced fodder production has increased the price of fodder and reduced cattle ownership in the region. This has also resulted in a decline of household dairy usage, with possible implications for household nutrition.

Borewells and indebtedness

There are about 250 drilling units in Coimbatore alone that charge around Rs. 200 per metre for drilling (in 2016). A traditional water diviner determines the exact location for the well. Farmers prefer to use water diviners, as they are cheaper and more easily available than hydrogeological experts. In 2012 the state government introduced the Tamil Nadu Minor Irrigation Scheme to provide expertise, machinery and funds for drilling wells.[7] However, lack of awareness, and an excessively bureaucratic and inefficient delivery system led many farmers to hire private contractors to drill borewells, obtaining the required funds through informal loans from moneylenders or by selling assets.

The costs associated with informal borrowing in the region range from 10 to 40 per cent (key informant interview). Tamil Nadu has some of the highest levels of household indebtedness in India at 82.5 per cent, compared to the national average of 52 per cent (NSSO 2013). Farm debt driven by social and economic aspirations is leading to pauperization, marginalization and growing arrogance on the part of the local elite (Guérin et al. 2011). Although the digging of borewells is not the sole reason for the high level of indebtedness in the region, it is strongly associated with reckless borrowing practices (Taylor 2013). In our study, 60–70 per cent of OBC farmers and close to 90 per cent of Pallar and Kurumba farmers reported having unpaid debts at the time of the interview in 2016 (see Table 7.1).

The hydrogeological characteristics of the region and the exploitation of groundwater have resulted in a high borewell failure rate. A study by Palanisamy et al. (2008) reported that farmers had a success rate of only 30 per cent in borewell drilling in Coimbatore district. A failed borewell can lead to financial ruin. Farmers often take loans in the hope that a bumper crop provisioned by irrigation will allow them to pay off debts. Depleting groundwater tables can drive farmers to bore up to three times in the same place to ensure a functioning well, often accruing substantial debts in the process. Around 67 per cent of the households in our study reported failed borewells as the primary cause of agricultural debt. The amount of debt varied significantly across castes, with SCs reporting a 24 per cent higher average amount of debt than OBC castes.

Rajeshwaree (42), a Pallar woman whose household owns 2.5 acres of land in Chinnakallipati village, described the indebtedness of her family due to multiple failed borewells on her farm: "I was not even aware of the cost of drilling the borewell, or the loan taken; we simply continue to pay monthly to the debt collector, hoping that one day we can clear our debts."

Rajeshwaree's household has accumulated debts amounting to almost Rs. 3.5 lakh with the local moneylender over the past 10 years. The annual income of the household from agriculture is Rs. 74,000. Unable to repay the full amount, they have resorted to taking smaller loans from multiple sources to clear the initial debt. Rajeshwaree's 17-year-old daughter had to discontinue her college studies and take up a job in a textile factory in

Tiruppur to help pay off some of the debt. Rajeshwaree currently works as a cleaner in the local panchayat office in addition to her domestic and farm duties. She also invests whatever small savings she manages to gather in the village-level self-help group and chit funds.

Indebtedness in the region has forced men and women to diversify their livelihood portfolios (Djurfeldt et al. 2008). 81.1 per cent of farmers belonging to the Pallar (SC) and 71.1 per cent belonging to the Kurumba (ST) communities practice seasonal migration. Around 74 per cent of the male farmers reported clearing farm debts as the main reason for seasonal migration. Indebtedness is disproportionally higher among small farmers belonging to SC (89.2 per cent) and ST (92.1 per cent) groups (Table 7.1), reinforcing the fact that coping strategies are shaped by class and caste positions. Seventy-two per cent of women above the age of 30 in our survey reported engaging in jobs outside the farm and the home, with 31 per cent engaging only in MGNREGA work, 40 per cent working as agricultural labourers and 29 per cent engaging in both. While SC (68.4 per cent) and ST (71 per cent) women engage in both MGNREGA and agricultural labour, women from OBC groups prefer not to work as agricultural labourers (24 per cent). Given the restrictions on their mobility due to domestic chores, and the limited options that married women have for earning additional incomes (Rao 2014), MGNREGA provides a valuable source of income for women during the summer months, compared to other physically demanding and poorly paid jobs available locally (Carswell and De Neve 2014).

Household cooperation, conflict and decision making

While women in rural South India are seen to have more autonomy and agency than their North Indian counterparts (Dyson and Moore 1983), they appear to have limited agency over water production (Zwarteveen 1997). Men maintain control over groundwater resources, from the decision to drill wells to well maintenance and water allocation. The drilling of boreholes is carried out by private male contractors who prefer to deal with the men of the household, perpetuating a masculine culture around borewells. As Zwarteveen (2008) notes, colonial and neocolonial relations of technology and gender have informed these 'traditional' (irrigators) and 'globalized' (engineers and mechanics) masculinities. This is despite the fact that women share the burden of work involved in irrigation, require water for livestock and household maintenance, and more importantly, often contribute to the cost of the borewells from their gold and other savings. Their exclusion here is justified by the social construction of the well as a male asset similar to other productive assets such as machinery or land across rural communities in India (Rao 2006; Deere et al. 2013).

Selvi (53) is a VokkaligaGounder woman whose household has 9 acres land in Chinnakallipatti. With eight failed borewells, she discussed the decision to invest in well boring:

The first well was dug about twenty years ago, before that we depended only on the rain. My husband wanted to start growing different types of crops, so he suggested that we dig the well … Initially, I was not convinced; it is a lot of money, and I was not sure if we would get water. But he convinced me and dug the well. After two years the water in the first bore-well decreased. We had also bought more land for agriculture and needed water to irrigate the crops; he decided to dig another well … My husband did not consult me on this …

> This bore failed, and I asked my husband to stop digging. He did not listen to me … If we had not spent so much money on the wells we could have invested in something else, maybe a petty *kadai* (small shop). Now we already have invested too much of our money in it, it is too late to do anything else.

Figure 7.3 shows how groundwater depth in Coimbatore district has fallen since 1999, with a dramatic dip in 2005. Selvi's narrative highlights the change in her attitude towards investing in borewells as the groundwater levels declined, and her resentment of her husband's decision to take higher risks by boring for more wells is palpable. Many women in the area expressed similar feelings, mentioning their limited agency in influencing decisions on the digging of borewells. Women, especially those in households with smaller landholdings, preferred to invest in small businesses or in assets such as gold and livestock; they regarded the digging of borewells as a precarious gamble that could result in financial ruin.

Borewells are useful to both men and women. However, as symbols of a productive farm and a successful (male) farmer, they are socially constructed as 'male' assets. During our interviews, borewell ownership was often used

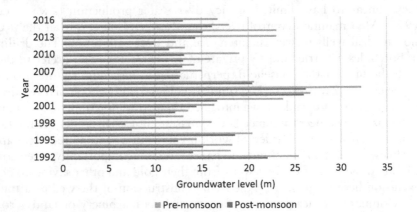

Figure 7.3 Average pre-monsoon (May) and post-monsoon (July) groundwater levels in Coimbatore District, 1997–2015.

Data source: Average annual groundwater levels in Coimbatore district (1997–2015) from Tamil Nadu Water Supply And Drainage Board (TWAD) data.

as a qualifier of success, even replacing land as a source of prestige and status within the community. Paramasivam (44), a VokkaligaGounderfarmer in Bhavani, had borrowed money from the village moneylender to dig a borewell four months previously, despite having one working borewell to irrigate his 3-acre field. When asked why he wanted to dig another well despite being aware of the risks, he pointed to his brother's adjacent field, stating "My brother has no worries, he has two working borewells and is doing very well in his farming and in his life". Competition in relation to well boring stemming from the associated feeling of success is shaping the behaviour of farmers in the region, often leading to indebtedness. Farmers like Paramasivam resort to borrowing from moneylenders due to the inflexibility of formal financial institutions over providing second loans where crop loans remain outstanding.

The construction of the borewell as associated with male identity can be unpacked using Sandra Harding's seminal text on how gendered social life is produced via symbolism (Harding 1986; Zwarteveen 2008). The hegemonic construction of irrigation, particularly borewell irrigation, can be attributed to the androcentric trappings of cash cropping, a co-production of irrigated systems in the region (Connell and Messerschmidt 2005). The very term 'cash crop', alluding to a crop grown only for sale, relegates it to an 'enterprise' that, under traditional cultural and societal norms, falls into the purview of men. In this case, as in similar cases elsewhere in Asia and in Africa, men control the growing and sales of cash crops such as banana, tobacco and curry leaf, while women retain control of the less remunerative fodder and food crops, even though they also contribute labour for cash crops (Rajamma 1993; Doss 2002). While women do not necessarily lack the knowledge or skills to engage with well boring, they have been systematically discouraged by both state institutions and the community from entering these 'technical', male-dominated spaces (Satyavathi et al. 2010).

Murugan, Selvi's husband,explained why he decided to bore multiple wells despite repeated well failures:

> I initially got the idea to dig a borewell after our *ooruthalevar* (village headman) dug his borewell. This was twenty years ago. It functioned well and allowed me to plant all the crops that I wanted. It provided enough water for the cattle and the house. But water in the well started decreasing in the past ten years. This is because everyone around has started to dig many wells. I had to ensure that I had enough water. So, I dug two more wells. These failed, so I took a loan and dug another one. Over the last eight years I have dug nearly five wells.... Of course, I have to keep digging to ensure that I have a functioning well. How else will I irrigate my crops? I cannot go back to rain-fed farming. All the farmers in the village have wells. If I have a good well I can make my farming very successful ... I cannot consult my wife regarding these decisions. She does not understand farming.

The borewells in Murugan and Selvi's farm have been financed through a combination of personal savings, bank loans and, more recently, loans from relatives and moneylenders. The household has sold assets to repay these loans, but they still have outstanding debts amounting to about Rs. 7 lakhs to pay off. Murugan hoped that he would at least be able to repay the moneylender after his next banana harvest.

The proliferation of borewells has had important implications for intra-household gender relations. Economic success and financial provisioning are an important conjugal expectation from men in patriarchal rural societies, who in turn expect their wives to meticulously perform familial and mothering roles (Rao 2012). Women may suggest alternative investment strategies, but once the borewell option is pursued, they have no choice but to also contribute to the repayment of loans. Men frequently resort to drinking to cope with a sense of hopelessness as a result of failed borewells and mounting debt burdens, sometimes even leading to violence. This link, however, requires further exploration (Rao 2015).

While tensions are rising due to failed borewells, new forms of cooperation are also visible. Although the primary impact of wells has been in the sphere of livelihoods, wells have also allowed easier access to water for domestic activities. Reproductive roles remain women's primary responsibility, and the direct burden of water provisioning falls on them. Yet, in the FGDs, women did indicate that when borewells failed, men increasingly contributed both cash and labour to the provisioning of domestic water for their households.

Savitree (32), aVokkaligaGounder woman from Marudur, with 4 acres of land in the household, discusses this shared responsibility:

> During the dry years, when there is less water in the borewell, my husband goes to the stand pipe [government-provided water tap in the village square] to collect water. I used to do it, but he has a bike, and it is easier for him to do it especially later in the day when it is not very safe for women to go out … At times of scarcity, we have to buy water from tankers for household needs like cooking and washing. My husband knows that I need water to cook for the family, so he will ensure we get water from the tanker even though it is expensive.

This is an interesting reversal of the gendered binary of domestic water as the domain of women, while its livelihood use becomes the domain of men (Van Koppen 2007; Zwarteveen and Ahmed 2012). Savitree and her husband Gopal both agree that it is no longer the responsibility only of women to collect water; if there is no water in the house, all of them will suffer. However, she does admit that during the summer months, when there is no work on the fields and her husband goes to the town for construction work, she has to collect the water herself.

Female-headed households

Feminist scholars have postulated that the atomization of water ownership through borewells allows women to circumvent historical inequalities in access to water by invalidating the necessity to participate in male-dominated patriarchal water user groups (Meinzen-Dick and Zwarteveen 1998; Ilahi and Grimard 2000). However, rights to water are tenuous and mediated through other constraining factors such as technology, knowledge and infrastructure (Boelens and Zwarteveen 2005; Ahlers and Zwarteveen 2009). This comes to prominence in the boring and maintenance process, when women have to depend on external support systems to operationalize their access to groundwater. Furthermore, groundwater access has never been fully equitable and is often determined by hydrogeological features of the land and shaped by patterns of marginalization propagated through, for instance, inheritance of less productive land (such as land located on hard rock aquifers).

There are a limited number of households in this region headed by widowed, divorced or unmarried women. The 2005 amendments to the Hindu Succession Act of 1956 established a gender-equal basis for inheritance of land and resources (Kelkar and Jha 2016). However, women continue to face a host of barriers and taboos that make their cultivation less productive than that of men (Agarwal 1994). Female-headed households usually have to diversify their livelihood activities and they prefer to undertake more 'gender-accepted' livelihoods such as animal husbandry and mill work.

Recounting her experience, Nagamma (48, VellalaGounder), a widow from Chinnakallipatti village who owns 5 acres of land, said:

> There was only one well on the 10 acres of land that my father owned. When it was time to divide the land, I had no say or choice on which land I received. I received an equal portion of land because it is the law, but my land does not have any water source. I have to buy water from my brother to irrigate the land but he usually does not have water to spare. I cannot dig a well as I do not have the money and I will not be able to do it by myself. I will require support from my brother, and I do not wish to be indebted to him. I have decided to sell the land as it has become unprofitable.

Nagamma cultivates only half of her land with finger millet and foxtail millet and has kept the rest fallow. She has limited access to water even for supplemental irrigation and usually suffers crop losses in dry years, particularly of finger millet, an important source of nutrition for the family, but which does not fare well in water-stressed conditions. The unprofitability of the land has made her decide to sell it. Nagamma has two daughters and is the sole earner for her family since her husband died 10 years ago. She works as an agricultural labourer and performs odd jobs (cleaning the local school) around the village to provide for her family as she cannot move out of the village for work. She says that she can save the money she gets from the sale of the land for her daughters' weddings. Nagamma is in a particularly vulnerable position, as she has no sons to help consolidate her social agency. She is vulnerable to land grabbing, mainly for real estate and house construction, from the local land mafia, and depends on her brother's family to provide security for her and her family.

Groundwater and gendered well-being

The development of groundwater in the region has been credited with increased farm profits, reduction of rural poverty and equitable access to water. Having access to a fully functioning borewell ensures uninterrupted water supply, allowing farmers to grow crops of their choice, increasing both productivity and incomes. Our research suggests that the availability of groundwater has had positive and negative implications for gendered well-being, varying across class and caste. Groundwater has allowed for the spread of intensive cash crop-based agrarian systems, augmenting livelihoods across social groups, and increasing gender wage parity for agricultural labourers, especially SC and ST women (see Table 7.2). It has also provided water for domestic usage, reducing women's domestic work burdens. Some fluidity can be observed in gender roles concerning water, with men taking increasing responsibility for domestic water provisioning and women spending more time in irrigation activities. However, with groundwater now reframed as a symbol of wealth and prosperity and not just a

Table 7.2 Intra-household strategies for coping with water shortages in order of preference

	OBC	SC	ST
Male	Selling livestock	Selling Livestock	Agricultural labor
	Deepening existing wells	Agricultural labor	Selling Livestock
	Buying water from tankers and neighbours		Collection of forest produce
	Temporary migration		
Female	Buying water from neighbours	Collecting water from public sources	Sourcing water for cattle from forested areas
	Selling livestock	Selling livestock	Agricultural labor
	Collecting water from public sources	Agricultural labor	Collection of forest produce

Note: The strategies are not mutually exclusive and vary within caste by age and social determinants such as marriage.
Source: Primary data from surveys, FGDs and interviews conducted in 2016.

productive asset, it has emerged as an embodiment of economic aspirations and successful masculinity.

Since 2011 this enhanced sense of well-being has begun to show signs of waning, with a decrease in the quality and quantity of groundwater. Farming households have indicated that household well-being, which in many ways is inextricably linked with groundwater, both in the domestic and the livelihood sphere, is seriously threatened by depletion of the resource in the region. Farmers in Chinnakallipatti reported that 20 years ago, groundwater was available at less than 200 feet. But now, they must dig to a minimum of 800 feet to access water. The depth of boring has increased, as has the number of failed borewells.

Failing borewells are severely affecting agricultural productivity and livelihoods, ensnaring communities in cycles of debt. Indebtedness has resulted in increased work burdens for both men and women. While men have wider options for more remunerative work owing to their mobility, women often have to shoulder additional burdens of paid and unpaid farm work.

Women invest more time in collecting water from other sources, such as public standpipes. In particularly dry months when water is bought from tankers, this expensive resource has to be strictly rationed. Also, with a focus primarily on cash crop cultivation, women's nutritional choices have been compromised.

Jayamala (62), a VellalaGounderwoman from Marudur village whose family has been farming in the region for the last three generations, talked about the present state of groundwater in the region:

> Twenty years ago, we used to only grow millets and sometimes vegetables. It was enough for the household. After we got access to a borewell, we have started growing jasmine, this gives us more money, but we have to buy our vegetables and grains from the ration shop or the village market ... The last five years we got less and less water from the wells. This year is the worst [2016]. Our crop has failed. We have no money to buy food, and there is no food in our fields.

Peer pressure and changing notions of masculinity, along with aspirations for more profitable farming and higher incomes to finance children's education and status-enhancing consumption, have led to competitive borewell digging. This has had implications for the financial security of the household, often giving rise to marital conflict. In this context, women have a tenuous hold on their assets, vital to their social standing, household agency and economic autonomy. In the absence of other risk-management strategies, women's assets are the first to be pawned or sold.

Despite the precarious state of groundwater in Tamil Nadu, there is no sign of the regulation of groundwater usage. A Groundwater Regulation Bill, which included provision for the setting up of the Tamil Nadu Groundwater Authority, was passed in 2003, but repealed in 2013. In 2015, efforts to regulate groundwater use were renewed and the Tamil Nadu Panchayats

(Regulation of Sinking of Wells and Safety Measures) Rules was passed. This made it mandatory for borewell operators to obtain permission from local panchayats before deepening and renovating borewells. However, enforcement of these rules is limited, and unregulated borewell digging and groundwater over-exploitation continues unabated in the region (Palanisami et al. 2008; Moench et al. 2012). Communal water sources, such as traditional tanks and farm ponds, have been neglected, leaving farmers with a lack of viable alternative water sources.

Although our entry point is groundwater, our analysis has allowed us to explore the complexities of decision making at the household level, where men and women hold gendered interests in resource management through their distinctive roles, responsibilities and livelihood stakes. These vary with class and caste positions, with women in smallholder households, mainly SC and ST, often bearing the brunt of growing indebtedness, higher work burdens and less nutritious food. Groundwater is 'multi-use' water, yet its primary purpose is seen as servicing livelihoods. This has allowed patriarchal norms and practices from the male-dominated agrarian space to be carried forward into decision making around the use of water. The experiences of women-headed households emphasizes the key gendered relational aspects of groundwater usage. Women continue to require the support of male family members to operationalize their access to groundwater, perpetuating the disadvantage they face in resource access, use and control in the agrarian sphere.

There is an urgent need to properly enforce regulation, not only to deal with the rapid depletion of groundwater due to variable monsoons and other developmental factors, but also to ensure that farming communities are not caught in a vicious cycle of indebtedness. This has particularly negative effects on women's status and well-being, as apart from intensifying labour burdens, indebtedness also contribute to a loss of women's assets, especially gold and livestock. Assetlessness, in turn, can enhance vulnerabilities to livelihood shocks, but equally to domestic violence. In rural communities, land and water are key livelihood assets, and their sustainable use becomes central to other forms of sustainability – livelihood security and social and gender equity. One needs to be mindful, then, that in dealing with water stress and drought, one does not intensify inequities of gender, class and caste.

Notes

1 Body of local self-government at the village level.
2 Scheduled Castes (SC), Scheduled Tribes (ST), Other Backward Castes (OBC). SCs largely cover communities have been historically disadvantaged and marginalized and are often associated with the socially degrading occupations. STs are characterized by geographical and social and economic isolation, and low economic status. OBC is a collective term used by the Government of India to classify castes which are educationally and socially disadvantaged. Source: Department of Social Justice and Empowerment, Government of India.

3 The MGNREGA is an Indian social security measure that guarantees 100 days of work per household per year.
4 Public Distribution System (PDS) is a scheme in India that facilitates the supply of essential food grains and commodities at a subsidized rate through a network of fair trade shops on a recurring basis. Source: Department of Food and Public Distribution
5 According to the National Sample Survey's 70th report, 'Situational Analysis of Agricultural Households 2012–13', the average income for farm households in Tamil Nadu is Rs. 7,000.
6 Paun is a unit of measurement of gold. It is equal to 8 grams, and valued at approximately Rs. 27,500 (July 2019).
7 http://www.tn.gov.in/scheme/data_view/19481 accessed on 7 July 2019.

References

Agarwal, B. 1994. *A field of one's own: Gender and land rights in South Asia* (Vol. 58). Cambridge: Cambridge University Press.

Ahlers, R. and M. Zwarteveen. 2009. 'The water question in feminism: Water control and gender inequities in a neo-liberal era', *Gender, Place and Culture*, 16(4): 409–426.

Anuradha, G. 2018. 'Public Distribution System in Tamil Nadu: Implications for Household Consumption.' *Leveraging Agriculture for Nutrition in South Asia (LANSA) Working Paper Series*, Volume 2018 Number 23. Chennai: MS Swaminathan Research Foundation (MSSRF).

Boelens, R. and M. Zwarteveen. 2005. 'Prices and politics in Andean water reforms', *Development and Change*, 36(4): 735–758.

Carswell, G. and G. De Neve. 2014. 'MGNREGA in Tamil Nadu: A story of success and transformation?', *Journal of Agrarian Change*, 14(4): 564–585.

Chinnasamy, P. and G. Agoramoorthy. 2015. 'Groundwater storage and depletion trends in Tamil Nadu State, India', *Water Resources Management*, 29(7): 2139–2152.

Connell, R.W. and J.W. Messerschmidt. 2005. 'Hegemonic masculinity: Rethinking the concept', *Gender & Society*, 19(6): 829–859.

Deere, C.D., A.D. Oduro, H. Swaminathan and C. Doss. 2013. 'Property rights and the gender distribution of wealth in Ecuador, Ghana and India', *The Journal of Economic Inequality*, 11(2): 249–265.

DES. 2000. *Season and Crop Report Tamil Nadu 1998–99*. Chennai: Department of Economics and Statistics, (DES), Government of Tamil Nadu.

DES. 2011. *Season and Crop Report Tamil Nadu 2009–10*. Chennai: Department of Economics and Statistics, (DES), Government of Tamil Nadu.

DES. 2013. *Season and Crop Report Tamil Nadu 2011–11*. Chennai: Department of Economics and Statistics, (DES), Government of Tamil Nadu.

Djurfeldt, G., V. Athreya, N. Jayakumar, A. Rajagopal, R. Vidyasagar and S. Lindberg. 2008. 'Agrarian change and social mobility in Tamil Nadu', *Economic and Political Weekly*, 43(45): 50–61.

Doss, C.R. 2002. 'Men's crops? Women's crops? The gender patterns of cropping in Ghana', *World Development*, 30(11): 1987–2000.

Dyson, T. and M. Moore. 1983. 'On kinship structure, female autonomy, and demographic behavior in India', *Population and Development Review*, 9(60): 35–60.

Gandhi, V.P. and N.V. Namboodiri. 2009. '*Groundwater irrigation in India: gains, costs and risks.*' Working Paper No. 2009-03-08. Ahmedabad: Indian Institute of Management.

Garikipati, S. 2009. 'Landless but not assetless: female agricultural labour on the road to better status, evidence from India', *The Journal of Peasant Studies*, 36(3): 517–545.

Geethalakshmi, V., A. Lakshmanan, D. Rajalakshmi, R. Jagannathan, G. Sridhar, A.P. Ramaraj, K. Bhuvaneswari, L. Gurusamy and R. Anbhazhagan. 2011. 'Climate change impact assessment and adaptation strategies to sustain rice production in Cauvery basin of Tamil Nadu', *Current Science*, 101(3): 342–347.

Gleeson, T., J. VanderSteen, M.A. Sophocleous, M. Taniguchi, W.M. Alley, D.M. Allen and Y. Zhou. 2010. 'Groundwater sustainability strategies', *Nature Geoscience*, 3(6): 378.

Government of Tamil Nadu (GoTN). (n.d.). *Report of the Animal Husbandry, Dairying and Fisheries Department (2007–2012)*. http://www.tn.gov.in/dear/ Animal%20husbandry.pdf

Guérin, I., R. Marc, G. Venkatasubramanian and S. Kumar. 2011. '*The social meaning of over-indebtedness and creditworthiness in the context of poor rural South Indian households (Tamil Nadu)*', RUME Working Papers Series 2011–1. Paris: IRD.

Harding, S.G. 1986. *The Science Question in Feminism*. Ithaca: Cornell University Press.

Heyer, J. (2016). 'Loosening the ties of patriarchy with agrarian transition in Coimbatore villages: 1981/2–2008/9.' In Mohanty, B. (ed.) *Critical Perspectives in Agrarian Transition: India in the Global Debate*. New Delhi: Routledge.

Ilahi, N. and F. Grimard. 2000. 'Public infrastructure and private costs: Water supply and time allocation of women in rural Pakistan', *Economic Development and Cultural Change*, 49(1): 45–75.

Janakarajan, S. 1999. 'Conflicts over the invisible resource in Tamil Nadu.', In Marcus Moench, Elisabeth Caspari and Ajaya Dixit (eds) *Rethinking the Mosaic: Investigations into Local Water Management*. Kathmandu: Nepal Water Conservation Foundation (NWCF) , pp. 123–158.

Janakarajan, S. and M. Moench. 2006. 'Are wells a potential threat to farmers' well-being? Case of deteriorating groundwater irrigation in Tamil Nadu', *Economic and Political Weekly*, 41(37): 3977–3987.

Kabirdoss, Y. 2017. 'Southern India reels under drought, Tamil Nadu worst hit', *Times of India*, 5 March. https://timesofindia.indiatimes.com/city/chennai/ south-in-drought-grip-water-in-tamil-nadu-dams-at-80-below-normal/article-show/57472927.cms. Accessed on 12 August 2019.

Kajisa, K., K. Palanisami and T. Sakurai. 2007. 'Effects on poverty and equity of the decline in collective tank irrigation management in Tamil Nadu, India', *Agricultural Economics*, 36(3): 347–362.

Kannan, E. 2013. 'Do farmers need free electricity?: Implications for groundwater use in South India', *Journal of Social and Economic Developmen*, 15(2): 16–28.

Kelkar, G. and S.K. Jha. 2016. 'Women's agential power in the political economy of agricultural land'. *Agrarian South: Journal of Political Economy*, 5(1): 98–122.

Kosalram, S.A. 1973. 'Political economy of agriculture in Tamil Nadu', *Social Scientist*: 1(12): 3–21.

Krishnaswamy, J., S. Vaidyanathan, B. Rajagopalan, M. Bonell, M. Sankaran, R.S. Bhalla and S. Badiger. 2015. 'Non-stationary and non-linear influence of ENSO and Indian Ocean Dipole on the variability of Indian monsoon rainfall and extreme rain events', *Climate Dynamics* 45(1–2): 175–184.

Kulkarni, S. 2016. 'Gender and water in India: A review', in V. Narain and A. Narayanamoorthy (eds) *Indian water policy at the crossroads: Resources, technology and reforms*. Heidelberg: Springer International Publishing, pp. 73–91.

Kulkarni, H., M. Shah and P.V. Shankar. 2015. 'Shaping the contours of groundwater governance in India', *Journal of Hydrology: Regional Studies*, 4: 172–192.

Kumar, M.D. and O.P. Singh. 2008. 'How serious are groundwater over-exploitation problems in India? A Fresh Investigation into an Old Issue.' *Managing water in the face of growing scarcity, inequity and declining returns: Exploring fresh approaches. 7th Annual Partners' meet of International Water Management Institute (IWMI)-Tata water policy research program*. Patancheru: International Crop Research Institute for the Semi-Arid Tropics (ICRISAT), pp. 2–4.

Livingston, M. 2009. *Deep wells and prudence: Towards pragmatic action for addressing ground water overexploitation in India*. Washington, DC: World Bank.

Meinzen-Dick, R. and M. Zwarteveen. 1998. 'Gendered participation in water management: Issues and illustrations from water users associations in South Asia', *Agriculture and Human Values*, 15(4): 337–345.

Moench, M., H. Kulkarni and J. Burke. 2012. 'Trends in local groundwater management institutions.' *Thematic paper 7*. Rome: Food and Agriculture Organization (FAO)

Mukherji, A., S. Rawat and T. Shah. 2013. 'Major insights from India's minor irrigation censuses: 1986-87 to 2006-07', *Economic and Political Weekly*, 48(26–27): 115–124.

Mukherji, A. and T. Shah. 2005. 'Groundwater socio-ecology and governance: A review of institutions and policies in selected countries', *Hydrogeology Journal*, 13(1): 328–345.

Narayanamoorthy, A. and R.S. Deshpande. 2003. 'Irrigation development and agricultural wages: An analysis across states', *Economic and Political Weekly*, 38(35): 3716–3722.

NSSO. (2013). *India—Situation Assessment Survey of Farmers, 2013, NSS 70th Round*. National Sample Survey Office (NSSO), Ministry of Statistics and Programme Implementation. New Delhi: Government of India.

Palanisami, K., A. Vidhyavathi and C.R. Ranganathan. 2008. 'Wells for welfare or illfare? Cost of groundwater depletion in Coimbatore, Tamil Nadu, India', *Water Policy*, 10(4): 391–407.

Phansalkar, S. and V. Kher. 2006. 'A decade of Maharashtra groundwater legislation: Analysis of the implementation process', *Law Environment and Development Journal*, 2(1): 69–83.

Prabu, P. and M.A. Dar. 2018a. 'Land-use/cover change in Coimbatore urban area (Tamil Nadu, India)—a remote sensing and GIS-based study.' *Environmental Monitoring and Assessment*, 190(8): 445.

Prabu, P. and M.A. Dar. 2018b. 'Land-use/cover change in Coimbatore urban area (Tamil Nadu, India)—a remote sensing and GIS-based study.' *Environmental Monitoring and Assessment*, 190(8): 445.

Promod, Madhav. 2017. 'CMO Pannerselvam declares entire Tamil Nadu drought hit, announces various schemes.' *India Today*, http://indiatoday.intoday.in/story/opanneerselvam-tamil-nadu-drought-hit-schemes-water-cauverywater/1/853909.html.

Puthiyasekar, C., M.A. Neelakantan and S. Poongothai. 2010. 'Heavy metal contamination in bore water due to industrial pollution and polluted and non- polluted sea water intrusion in Thoothukudi and Tirunelveli of South Tamil Nadu, India', *Bulletin of Environmental Contamination and Toxicology*, 85(6): 598–601.

Rajamma, G. 1993. 'Changing from subsistence to cash cropping: Sakaramma's story', *Gender & Development*, 1(3): 19–21.

Rao, N. 2006. 'Land rights, gender equality and household food security: Exploring the conceptual links in the case of India', *Food Policy*, 31(2): 180–193.

Rao, N. 2012. 'Breadwinners and homemakers: Migration and changing conjugal expectations in rural Bangladesh', *Journal of Development Studies*, 48(1): 26–40.

Rao, N. 2014. 'Caste, kinship and life-course: Rethinking women's work and agency in rural South India', *Feminist Economics*, 20(4): 78–102.

Rao, N. 2015. 'Marriage, violence, and choice: Understanding dalit women's agency in rural Tamil Nadu', *Gender and Society*, 29(3): 410–433.

Rao, N. 2017. 'Assets, agency and legitimacy: Towards a relational understanding of gender equality policy and practice', *World Development*, 95: 43–54.

Roy, A.D. and T. Shah. 2002. 'Socio-ecology of groundwater irrigation in India', in Ramón Llamas, Emilio Custodio (eds) *Intensive use of groundwater. Challenges and opportunities*. Lisse, The Netherlands: A. A. Balkema, pp. 307–335.

Saravanakumar, K. and R. Ranjith Kumar. 2011. 'Analysis of water quality parameters of groundwater near Ambattur industrial area, Tamil Nadu, India', *Indian Journal of Science and Technology*, 4(5): 660–662.

Satyavathi, C.T., C. Bharadwaj and P.S. Brahmanand. 2010. 'Role of farm women in agriculture: Lessons learned', *Gender, Technology and Development*, 14(3): 441–449.

Schlink, A.C., M.L. Nguyen and G.J. Viljoen. 2010. 'Water requirements for livestock production: A global perspective', *Revue Scientifique et Technique*, 29(3): 603–619.

Sekhri, S. 2014. 'Wells, water, and welfare: The impact of access to groundwater on rural poverty and conflict', *American Economic Journal: Applied Economics*, 6(3): 76–102.

Selvaraj, K.N. and C. Ramasamy. 2006. 'Drought, agricultural risk and rural income: Case of a water limiting rice production environment, Tamil Nadu', *Economic and Political Weekly*, 41(26): 2739–2746.

Sen, A. 1990. 'Gender and cooperative conflicts', in I. Tinker (ed.), *Persistent inequalities: Women and world development*. New York: Oxford University Press, pp. 123–149.

Shankar, P.S.V., H. Kulkarni and S. Krishnan. 2011. 'India's groundwater challenge and the way forward', *Economic and Political Weekly*, 46(2): 37–45.

Swaminathan, H., R. Lahoti and J.Y. Suchitra. 2012. 'Gender asset and wealth gaps: Evidence from Karnataka', *Economic and Political Weekly*, 47(35): 59–67.

Taylor, M. 2013. 'Liquid Debts: Credit, groundwater and the social ecology of agrarian distress in Andhra Pradesh, India', *Third World Quarterly*, 34(4): 691–709.

Upadhyay, B. 2005. 'Women and natural resource management: Illustrations from India and Nepal', *Natural Resources Forum*, 29(3): 224–232. Oxford, UK: Blackwell Publishing Ltd.

Van Koppen, B. 2007. 'Dispossession at the interface of community-based water law and permit systems.' In Barbara Van Koppen et al. (eds) *Community-based water law and water resource management reform in developing countries*. Wallingford: Center for Bioscience and Agriculture International (CABI), pp. 46–64.

Walker, B.H. 2012. *Management of semi-arid ecosystems*. Amsterdam: Elsevier, pp. 245–250.

Zwarteveen, M. and S. Ahmed. 2012. 'Gender, water and agrarian change: an introduction'. In Margreet Zwarteveen et al. (eds) *Diverting the flow: Gender equity and water in South Asia*. New Delhi: Zubaan, pp. 303–311.

Zwarteveen, M. 2008. 'Men, masculinities and water powers in irrigation', *Water Alternatives*, 1(1): 111.

Zwarteveen, M. 1997. 'Water: From basic need to commodity: A discussion on gender and water rights in the context of irrigation', *World Development*, 25(8): 1335–1349.

8 Gender, migration and environmental change in the Ganges-Brahmaputra-Meghna delta in Bangladesh[1]

Katharine Vincent, Ricardo Safra de Campos, Attila N. Lázár, and Anwara Begum

Introduction

The relationship between gender and migration is complex. Gender differences in roles and relations affect who migrates, to where and with what purpose. Migration also has gendered effects in both receiving and sending areas and, in turn, affects roles and relations. Migration has long been an important component of livelihood strategies in Bangladesh. The majority of migration is from rural areas due to lack of employment, as has been recognized in the Sixth and Seventh Five Year Plans. Migration to urban areas is one of the major contributors to urban growth (Afsar 2003; Government of Bangladesh 2012, 2015).

Bangladesh is situated at the confluence of three major rivers – the Ganges, Brahmaputra and Meghna – with a large proportion of its surface area defined as a delta where these rivers meet the Bay of Bengal. Deltas are characterized by change, with the land areas constantly shaped and reshaped by the interaction of river sediment loads from upstream and the erosive capacity of the sea. Populations living in deltas have become accustomed to this variability. However, the rate of environmental change is exacerbated by human activities and climate change, which alter the context in which livelihoods are carved out and in which migration decisions are made (Szabo et al. 2015).

In this chapter we investigate the linkages between gender, migration and environmental change. We first outline some of the theory on migration and gender, on how environmental change affects migration, and the gendered effects of environmental migration. Then we illustrate migration patterns

1 We gratefully acknowledge all those involved in the design and implementation of the survey (enumerators and participants) on which this chapter is based, as well as Shouvik Das for producing Figure 8.1. This work is carried out under the Deltas, Vulnerability and Climate Change: Migration and Adaptation (DECCMA) project (IDRC 107642) under the Collaborative Adaptation Research Initiative in Africa and Asia (CARIAA) with financial support from the UK Government's Department for International Development (DfID) and the International Development Research Centre (IDRC), Canada. The views expressed in this work are those of the creators and do not necessarily represent those of DfID and IDRC or its Board of Governors.

in Bangladesh, drawing on primary data from a sex-disaggregated survey of 1356 households in the Ganges-Brahmaputra-Meghna delta conducted between 2015-16. Next we present some of the effects of these patterns on migrant-sending areas, and then finally we conclude with some insights into future environmental change and migration.

Unpacking migration

Migration is a complex phenomenon with varying dimensions and definitions. As a result, there are multiple theories explaining its drivers and decisions, and the range of intertwined factors associated with migration and other forms of human mobility (Bakewell 2010; Massey et al. 1993). Migration is often characterized by its duration, with permanent, seasonal and circular being common distinctions. Hugo (1978) embraces this by highlighting the idea of commitment within migration, in that there will be a change in "usual place of residence" and a sense of "permanence" in the move only if it is followed by a change in commitment from one area to another. Considering commitment can therefore shed light on various long-term movements (e.g., return or permanent migration) and the changing circumstances in which they occur (Begum 1999).

Reasons for migration also differ. The New Economics of Labour Migration theory posits that migration is economically driven and occurs to increase income (Castles 2010; Stark and Bloom 1985). The structural-functionalist approach emphasizes the role of the wider socio-economic and environmental structure in which individual action is embedded (Gerold-Scheepers and van Binsbergen 1978). Rational choice theories and the methodological individualist approach highlight that people's migration decisions reflect the nature of their human capital and are designed to be income maximizing (Haug 2008; Todaro 1969, 1976). Although the nature of the motivation may differ, whichever theory is applied the unifying thread is the perception that migration will bring improvements to livelihoods and wellbeing, and are thus associated with development (Bakewell 2010; Massey et al. 1993).

Gendered migration patterns

Many countries, including Bangladesh, collect sex-disaggregated data for internal migration, which enables interrogation of differences in patterns between women and men. Historically marriage is a major cause of migration, particularly for women (GIZ-RMMRU 2014). The extent of women's migration as related to marriage reflects gender roles and relations. Gender roles are the prescribed behaviours and characteristics that are allowed, expected and valued of men and women (Alston 2013; Begum 2015). They are socially and culturally defined, and so vary from place to place, but also over time. Broadly speaking, men are typically associated with 'productive' tasks, i.e., those that generate economic income, whilst women are typically associated with 'reproductive' tasks, i.e., care tasks related to the house-hold, and looking after children and the elderly (Islam 2014). Since gender roles dictate that earning income is typically the responsibility of men, much labour migration has traditionally been of men within the informal sector.

Despite migration trends being typically skewed in favour of men, a 'feminization of migration' is now occurring, both within Bangladesh and internationally (Castles and Miller 1993; Hossain et al. 2013). The increase in female migration has been attributed to various factors relating to the receiving area (pull factors) and the sending area (push factors) (Willis and Yeoh 2000). Growing access to education and level of education is a pre-requisite for women's participation in the labour market (Williams 2009). The lower costs of international migration and changing policies that sup-port it have also encouraged female out-migration (Government of Bangla-desh 2015). That said, gender roles and relations often still impede ability to migrate, even when the intention is there, particularly for women with lower levels of education. Based on a large panel survey, using individu-al-level data from 148 countries between 2009 and 2013, women are more likely to want to migrate if they do not feel treated with dignity and respect. However, the likelihood that these intentions are acted upon reflects income and family obligations, as well as the extent of their networks (Ruyssen and Salomone 2018).

Determining the causes of migration of men and women is inextrica-bly linked with gender roles and relations. Whilst female migration may be increasing, and this is associated with better education and economic labour opportunities, these may still be different from those of men. Large worldwide flows of women to do paid domestic work, for example, are fed by inequalities of gender, race, class and nationality, and in turn con-tribute to reproducing these differences (Altman and Pannell 2012). The links between gender and migration can be understood within a broader framework of social change: one in which gender roles and relations are considered (Lutz 2010). Gender differences in migration intersect with other social identifiers which contribute to power relations. In Bangla-desh, migration is reconfiguring class and gender identities but it is also mediated by religion and the modern Islamic identity (Rao 2014). Whilst women are often left behind, patterns of migration are also linked to the construction of masculinities, with pressures to provide for the family in a globalizing world linked to transnational mobility (Yeoh and Ramdas 2014).

Environmental change and migration

Migration has long been one of the many ways in which humans adapt to environmental change. Increasing concerns about growing environ-mental problems have augmented the interest of the scientific community and policy makers in the linkages between population movements and the environment (Black et al. 2011; Hugo 2011). Furthermore, in the past climate-induced displacement and environmental migration were addressed separately from mainstream migration theories. This approach is now considered antiquated among researchers and practitioners, and in policy circles, with environmental change, migration and development increasingly being examined together (Bettini and Gioli 2016).

The role of the environment in migration plays out in many circumstances. In Bangladesh, for example, there is a long-established seasonal pattern of migration from rural areas, with rural dwellers seeking work in urban areas (Etzold et al. 2014). This occurs during the lean agricultural season, or during flooding, particularly in the *Monga*-prone districts in the north-west and the north-eastern *Haor*-affected areas (Afsar and Baker 1999; Hossain et al. 2003; Marshall and Rahman 2012). Who migrates in this way and where is contingent upon availability of capital to cover the costs of such migration (McLeman and Smit 2006). Moving out from areas linked to environmental change suggests that deltas will experience net out-migration – although in Bangladesh the second-largest city, Chattogram (Chittagong), is within the delta, meaning that rural–urban migration, even to the largest urban areas, can also take place within the delta.

With climate change and climate change-related environmental stress, it is the likely magnitude and permanence of migration, not to mention its involuntary nature, that will be different. Strong policy discourses at international level talk of the likely high numbers of 'climate migrants' and 'climate refugees'. According to some estimates, there will be 200 million migrants by 2050 as a result of climate-induced push factors (including sea-level rise, coastal erosion, land degradation, floods, etc.) (Myers 2005; Stern Review Team 2006), although others suggest that migration may be even more significant (Black et al. 2011; Foresight 2011). However, the limited empirical evidence on this relationship has generated highly politicized debates. Estimates of the number of environmentally driven migrants, present and future, remain piecemeal, with considerable variation in projected figures (Gemenne 2011).

The emotive consequences of such mobility for human security, as well as the political implications, have constructed a discourse based on the intangible nature of climate migrants. It has been posited that such climate migrants are more of a "hyperobject" than a reality (Baldwin 2017). This is because, despite evidence of climate change-related pressures, the environment is rarely cited as the main reason for migration. In the Maldives, more than 50 per cent of respondents perceive future sea-level rise as a serious challenge at national level, but at individual level cultural, religious, economic and social factors play the most important roles in making decisions to migrate or not (Stojanov et al. 2016). Even in the face of exposure to climate change, for example floods or sea-level rise, whether or not people migrate is contingent upon a variety of factors affecting their vulnerability (Perch-Nielsen et al. 2008).

The causes and nature of environmental migration have variously been investigated, which can give some insights as to whether migration is a failure of adaptation, maladaptation or whether it is an effective adaptation response (Agrawal 2010). The roles of environmental degradation, water availability and extreme events have been researched. Exposure to extreme events such as floods, and loss of assets due to such events, can create large numbers of displaced people (Laczko and Aghazarm 2009). Extreme events have also been shown to be a driver of migration in Bangladesh (Joarder and Miller 2013). However, when migration is planned and voluntary, it can provide the resources necessary to improve wellbeing in the sending households (Adger et al. 2003; Tacoli 2009). Hence migration is more likely to be an adaptation (as opposed to a failure of adaptation) when it is planned and supported, and not forced and occurring under distress (Kartiki 2011).

Gendered effects of environmental migration

The relationship between environment and migration is gendered in terms of who moves, but also in terms of the implications. Increasingly the adaptation, resilience and vulnerability literature is engaging with gender; when considered through a gender lens, differences between men and women emerge (Bunce and Ford 2015). Whether or not migration acts as an adaptation differs between men and women, and it also reflects (and reinforces) existing gender relations. Social relations are altered as a result of climate change (Pearse 2017; Rao and Mitra 2013). To a certain extent, the effect on women depends on whether there has been forced or a planned and voluntary migration (Biao 2007), otherwise known as "erosive" or "content" migration (Warner and Afifi 2014). If the migration is voluntary, there is the possibility that ongoing interactions with the migrant may increase economic wellbeing (e.g., through remittances) and also social wellbeing. *De facto* female-headed households, where a woman is the head as a result of migration, often benefit from differently defined gender roles. This means that these women may be afforded access to community decision-making structures, for example, whereas if their husband were at home he would preferentially participate. However, if the migration has been forced or undertaken in conditions of distress, this may not be the case, and there is the risk that vulnerability may be exacerbated. In India, where tradition is for the wife of a migrant to join her in-laws in his absence, there is evidence that women's economic and social mobility is impeded (Desai and Banerji 2008; Gulati 1993).

Taking an intersectional approach also illuminates how other identifiers intersect with gender to determine whether migration improves or erodes the adaptive capacity of women left behind. Marital status plays a more significant role in affecting the adaptive capacity of women compared to men. In a study of the adaptive capacity of single, married, divorced and widowed men and women in Tanzania, widows and female divorcees were found to be relatively more disadvantaged in the field of agricultural water management, whilst divorced women assumed relatively more income-generating activities outside the farming sector (Van Aelst and Holvoet 2016). In tropical delta regions, out-migration is contributing to changing population structures and increasingly ageing populations, which has implications for the nature of vulnerability (Szabo et al. 2016). Thus, intersectionality is also important in determining the role of migration in social relations and the implications for adaptive capacity.

Investigating migration through a household survey

Fifty locations in the Ganges-Brahmaputra-Meghna delta were selected for implementation of the cross-sectional household survey. Sampling was undertaken in two stages. The first stage divided the study areas into five hazard zones (very low, low, medium, high, very high). Each cluster of households in the study area was assigned one of five hazard categories based on the risk category with the greatest percentage coverage in the cluster. In each stratum (multi-hazard zone), a number of clusters were selected proportional to the number of clusters in each stratum. Once clusters had been selected, a household listing took place based on demographic and migration characteristics. This approach resulted in 1356 valid sex-disaggregated questionnaires across 14 districts: Bagerhat, Barguna, Bhola, Chandpur, Chattogram, Cox's Bazar, Gopalganj, Jessore, Khulna, Laxmipur, Noakhali, Pirojpur, Potuakhali and Satkhira. The survey captures voluntary and planned migration more than forced displacement, and thus the results presented are biased towards these migration types, with supplementary information on displacement where available.

Migration patterns in Bangladesh

Who migrates, where, why and for how long?

The migration patterns highlight who migrates where and our survey data provides insights into the reasons for, and duration of, the migration. The vast majority of migrants are men. Survey findings reinforce earlier studies that found that 75 per cent and 83 per cent of internal migrants are male (Hossain et al. 2013; Sharma and Zaman 2009). They migrate internally, typically from rural areas to urban areas, which confirms findings from earlier panel surveys (e.g., Rahman et al. 1996) (Figure 8.1). Dhaka, which represents nearly 40 per cent of the urban population of Bangladesh, had net annual migration arrivals during the 2000–10 period of 300,000–400,000 (teLintelo et al. 2018). Approximately 58 per cent of all internal migrants in Bangladesh head to Dhaka (Siddiqui and Mahmud 2014). Chattogram grew by 3.6 per cent per year across the period 1990–2011 and had an aggregate population of four million in 2011 compared to 3.3 million in 2001 (Mia et al. 2015). Analysis of the household survey data revealed that Dhaka and Chattogram together accounted for almost 43 per cent of all

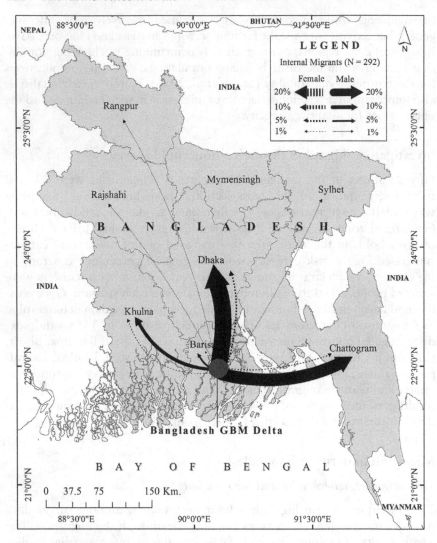

Figure 8.1 Gendered migration patterns from the Ganges-Brahmaputra-Meghna delta in Bangladesh.

Source: Shouvik Das using DECCMA survey data, 2016–17.

internal migration. Only 8 per cent of internal migrants in the survey were women, and over half of them moved to Dhaka, with smaller numbers migrating to Chattogram and Khulna. Government data also suggest that both male and female migration is increasing over time (BBS 2008).

Although the majority of migrants in our survey were internal, moving within Bangladesh's borders, a significant number also migrated internationally. Trends for international migration from Bangladesh have existed since independence in 1971 and patterns reflect colonial history.

A considerable number of Bangladeshis have been living abroad for some time, relocating primarily to the United Kingdom (Hadi 2001). In 2004, the size of the Bangladeshi diaspora in selected developed countries was around 1.5 million (Ministry of Expatriates' Welfare and Overseas Employment and IOM 2004). The findings of the household survey revealed that 37 per cent of migrants have relocated to international destinations. Oman (26 per cent), United Arab Emirates (24 per cent) and Saudi Arabia (16 per cent) were the most prevalent destinations among Bangladeshi migrants. Malaysia (7.5 per cent), Bahrain (5.8 per cent), Qatar (5.2 per cent) and Kuwait (4.0 per cent) were also cited by survey respondents.

The vast majority of international migrants are male. In our survey less than 3 per cent of the 173 reported international migrants were females. However, like female internal migration, female international migration is increasing. In 2006 the Bangladesh government lifted a restriction on female workers and since then the number of women migrating abroad has increased significantly. However, numbers are still small compared to men due to gender roles and relations creating an expectations that women should stay in the home.

For both internal and international migration, economic opportunities are the major reason for migration. Within Bangladesh, urban areas provide opportunities in both formal and informal sectors of the economy associated with the rapid industrialization and urban growth of the country. The majority of internal migrants captured in the household survey are waged workers (36 per cent) or own a small, mostly informal, business (9 per cent). The same is true for international migration, which has grown as a result of the ongoing fragile socio-economic conditions in Bangladesh that have driven people to seek employment opportunities elsewhere. The Middle East and Asia provide many opportunities for unskilled and semi-skilled labour, for example in the construction industry.

Figure 8.2 shows that, among the 156 households with a migrant, economic reasons for migration are by far the most prevalent, cited by nearly two-thirds of migrant families. This category includes pull factors, such as employment, and push factors, such as debt or loss of income in one or more seasons. The second most common motivation for migration is family reasons, for example a woman moving either to join her husband or for marriage. Nevertheless, these account for less than one-fifth of migrant households. Environmental stresses, such as environmental degradation and extreme events, account for 13 per cent of migration. This is consistent with an earlier study where 10 per cent of respondents attributed the primary reason for their migration to climatic stresses, although many of them had experienced different types of climate stresses and reported challenges of food security (Martin et al. 2013). Instead over a third said that their motivation for leaving was to earn a better income and have a better life. Housing problems, social/political problems and education each account for less than 5 per cent. The reason for migration is linked with gendered patterns. The gender roles and expectations are that men should engage in

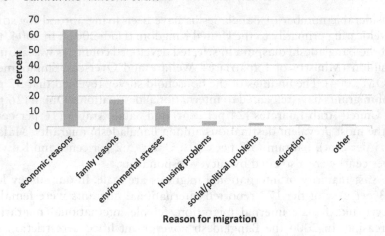

Figure 8.2 Reason for migration in migrant households (n = 463).

Source: DECCMA survey data, 2016–17.

productive labour to support their families, so when there is a need to move for economic reasons, it is more likely that the man will be the one to leave.

The duration of migration varies. Overall in Bangladesh, there is an increasing trend for temporary and circular migration, with short-term durations dominating over long-term ones (Hossain et al. 2013). The over-whelming majority (84 per cent) of international migrants are employed under a temporary contract that usually lasts three years.

Displacement

The migration outlined above is voluntary, in that people made the choice to leave. Forced migration, or displacement, whereby people have to move from their homes, is also common (Begum 2017). Situated on a delta, with the majority of the land less than 10m above mean sea level, Bangladesh is exposed to cyclones, flooding and erosion. Annually approximately 60,000 people are made landless as a result of the normal reshaping of the *chars* due to erosion (Hutton and Haque 2003). On top of this, an estimated 50 million people are exposed to, or affected by, disasters every five years, with the coasts facing a severe cyclone every three years on average, and a quarter of the country getting inundated during the annual monsoon rains (Shamsuddoha et al. 2013). Loss of houses and/or land can lead to forced migration or displacement.

Forced migration as a result of displacement does not exhibit the same gender differences as voluntary migration, nor the same patterns. Post-disaster movement typically involves short distances, but large numbers of people, with whole families moving together. For example, the 1998 floods inundated 61 per cent of the country, rendering 45 million people homeless. There were 26 major cyclones from 1970 to 2009, with the largest, Sidr, in

2007, displacing 650,000 people. Cyclone Sidr left US$1.7 billion of damage, approximately half of which was attributed to houses lost in the storm (World Bank 2010).

Effects of migration on migrant-sending areas

Voluntary migration rarely involves a whole family moving at one time. Instead it is more likely that one person from a household, typically a man, will migrate, leaving behind other members of the family in the migrant-sending area. The consequences of migration, therefore, exist not only for the migrant but also for the family (s)he leaves behind. In this section we outline some of the effects of migration in migrant-sending areas, and the extent to which migration is deemed successful.

One of the most prominent positive effects of migration on migrant-sending areas is the receipt of remittances. Remittances refer to money and/or goods that are sent back home by the migrant. Since the bulk of migration is driven by economic reasons, it is not surprising that flows of money and/or goods take place. Figure 8.3 shows that over 84 per cent of migrant households receive money and/or goods from their migrant members. Money is remitted most frequently, usually weekly or every 4–5 weeks, with monthly amounts equivalent to up to 50,000 Bangladesh Taka (approximately $550).

The receipt of remittances improves the economic status of the family left behind (Szabo et al. 2018). This can have impacts in terms of poverty reduction and can enable improvements in food security, health and education status, and housing. It can also enable purchase of productive assets. In a comparison of landholding sizes, the average for (internal) migrant households was 114.3 decimals, compared with 73.9 decimals for non-migrant households (Hossain et al. 2013). In our survey, the bulk of remittances is

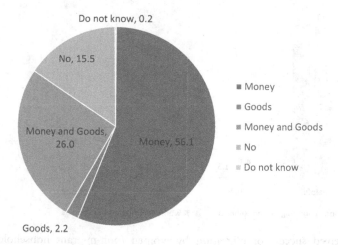

Figure 8.3 Percentage of migrant households receiving remittances (n=463).

Source: DECCMA survey data, 2016–17.

spent on daily necessities (e.g., food). The economic impacts of international migration are often more significant, due to the potential for higher earnings than internal migrants are able to achieve. There is also positive evidence of the impacts of women's migration in terms of contribution to family income. In one study, almost half of the women migrating abroad became the principal income earner in the household during their period of migration (Siddiqui 2001).

Remittances and the economic benefits of migration are likely to be a key component of the perceived success of migration. Figure 8.4 highlights that over 89 per cent of women in both non-migrant and migrant households deem migration to be helpful. This suggests that the material benefits are evident even to those households that are not yet participating in migration. Less than 10 per cent see it as unhelpful, neither or are not sure.

Whilst remittances are a positive effect of migration, the social effects and implications for gender roles and relations can be negative. In some cases, women report greater autonomy and decision-making capacity when their husband has migrated, as society is more flexible with social norms and more tolerant of women's involvement in what are often seen as men's tasks. On the other hand, migration can increase burdens on women left behind due to their reproductive care roles (Hossain et al. 2013). This is often exacerbated in the multi-generational households that are typical of rural Bangladesh. A woman typically moves in with her husband's family at marriage and if he then migrates, she has sole responsibility for the daily support of her parents-in-law. There can be a significant loss of autonomy in this case, as the mother-in-law takes priority in decision making. Our survey also included questions about levels of wellbeing. Overall, over two-thirds of women deem themselves happy. However, slightly fewer women in

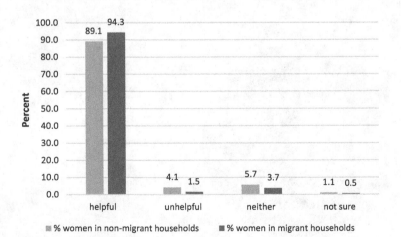

Figure 8.4 Perceived success of migration by women (non-migrant households n=844; migrant households n=404).

Source: DECCMA survey data, 2016–17.

migrant households deem themselves happy and slightly more deem themselves unhappy than women in non-migrant households (Figure 8.5).

Often, migration is an unfolding process as opposed to a one-off event. Typically, a man will migrate first alone, in order to establish himself in the receiving area in terms of employment and housing. If this process is successful, his family will often join him at a later date. In our survey nearly two-thirds of the women in migrant households expressed their intention to migrate in the future (Figure 8.6). However, migration is not an easy

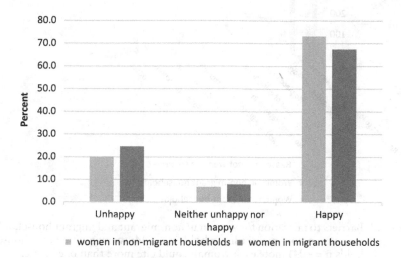

Figure 8.5 Level of happiness of women in migrant and non-migrant households (women in non-migrant households n = 844; women in migrant households n = 404).

Source: DECCMA survey data, 2016–17.

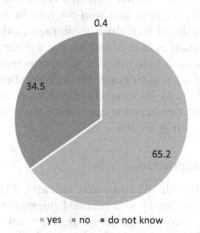

Figure 8.6 Percentage of women in migrant households who intend to migrate in the future (women in migrant households n = 404).

Source: DECCMA survey data, 2016–17.

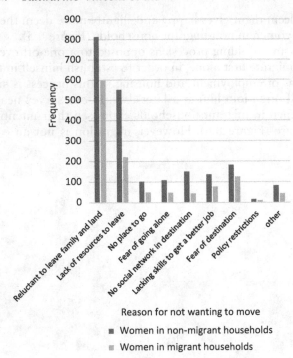

Figure 8.7 Barriers to migration for women in non-migrant and migrant households (women in non-migrant households n = 844; women in migrant households n = 404) (note each woman could cite more than one barrier).

Source: DECCMA survey data, 2016–17.

decision, and there are often significant ties of attachment to place that have to be overcome in order to move. Figure 8.7 shows that this is the major barrier to migration of women in both migrant and non-migrant households. However, for women whose husband has already migrated, other barriers relating to the receiving area, such as fear of not having a place to stay, not having a social network there or it being too crowded/expensive/dangerous are lower than for women in non-migrant households. Being part of a migration network, where there is information about work and people who can support integration into the receiving area, has been found to be important in any migration (Hossain et al. 2013).

Future environmental change and migration

Migration has a long history in Bangladesh (Bhuiyan and Siddiqui 2015; Siddiqui 2003). The vast majority of voluntary and planned migration is internal and from rural to urban areas, driven by economic reasons and the opportunities for employment. With Dhaka outside the delta and Chattogram within it, some migrants are moving outside the delta, whereas others are changing location within it.

Historically men have been, and continue to be, far more likely to migrate than women, although migration of women is increasing over time, including international migration. Whilst whole families may be displaced and forced to migrate, voluntary and planned migration typically involves selected members of the family – at least initially. Many women left behind state their intentions to migrate in the future when financial circumstances allow. Since it is more common for selected members to migrate, and typically men, the migration process has consequences for the households left behind and, in particular, for women.

Migration changes the population structure of sending areas, and has gendered consequences. When remittances are sent, this plays a role in increasing economic wellbeing. Gender roles are often more fluid as a result of migration, which can benefit or disadvantage women, depending on their circumstances. The psycho-social consequences of migration and the splitting of families can lead to different levels of happiness in migrant compared to non-migrant households.

The role of environmental factors in migration patterns is already being discerned, and is likely to continue into the future. However, since much of the country is a delta, it is difficult to determine delta-specific findings, and this explains why analysis of the survey reinforces existing understanding of the drivers and patterns of migration. However, these patterns may change in the future.

The geography of Bangladesh, as a low-lying coastal delta country, means that in addition to changes in temperature and rainfall patterns, it will be exposed to other climate changes such as sea-level rise. The amount of the country that will be flooded under a global average temperature increase of 3°C is more than 2.5 times that which would be flooded if the aspiration of limiting temperature increase to 1.5°C is achieved (Brown et al. 2018). Migration linked to climate change – although not solely for adaptation – is likely to increase (Saul 2012; Stojanov et al. 2016). In particular, the level of displacement is projected to increase. By 2050, it is estimated that over 35 million will be displaced in Bangladesh as a result of climate change – and this will lead to increased internal and external migration (DSK 2012). A consequence of the changing environmental conditions will be a change in the context in which migration decisions are made, with effects on migration patterns (Call et al. 2017). Although very few people cite the primary reason for migrating as environmental or climate – related, in many cases environmental and climate change are contributing to the stresses on livelihoods that prompt migration (Dasgupta et al. 2014).

The effectiveness of adaptation will play a key role in determining migration patterns and consequences. Structural adaptations, such as dikes and polders, are necessary to protect delta land and reduce the risk of large-scale displacement. *In-situ* livelihood adaptations enable people to earn a sustainable living in the delta, meaning that if they do take the decision to migrate it is likely to be planned and voluntary rather than forced. The government of Bangladesh has signalled its commitment to adapting to climate change through the National Adaptation Programme of Action and forth-coming National Adaptation Plan, the Bangladesh Climate Change Strategy and Action Plan, and the Bangladesh Climate Change Trust Fund. How-ever, establishing effective linkages between national and local levels can be challenging (Stott and Huq 2014). The implementation of these policies is uneven, with implications for adaptation at local level (e.g., Mahmud and Prowse 2012). At the same time, there have been critiques of the extent to which national policies are informed by local realities (DSK 2012; Haque et al. 2012). Whilst some infrastructural adaptation decisions need to be taken at district or national level, there is a strong trend for communi-ty-based adaptation in Bangladesh, and supporting the development of social institutions can reduce adverse impacts of climate change (Ali et al. 2014). Gender-sensitive adaptations will be essential to ensure women have equitable access to livelihood options that afford them choices and ensure that if they do move it is out of choice, and not forced.

At the same time, planning for future migration flows will be essential. Currently the greatest risks from climate displacement come from human insecurities that result from precarious alternative homes, for example in the growth of urban slums, and from social conflict (DSK 2012). Interviews with women migrants in urban slums in Chattogram highlighted that they are often subject to particular exploitation (Ava et al. 2017). There is also a need to develop appropriate legislation to protect displaced persons. Ref-ugees who cross international borders are recognized and have legal rights but, as yet, no similar legislation exists to protect the human rights of inter-nal migrants (Allan et al. 2016). Putting such frameworks in place is likely to contribute to the attainment of gender equality in Bangladesh.

References

Adger, W.N., Huq, S., Brown, K., Conway, D. and Hulme, M. 2003. 'Adaptation to climate change in the developing world', *Progress in Development Studies*, 3(3): 179–195.

Afsar, R. 2003. 'Internal migration and the development nexus: The case of Bangladesh', Presented at the *Regional Conference on Migration, Development and Pro-Poor Policy Choices in Asia, Refugee and Migratory Movements Research Unit and DfID*, Dhaka, Bangladesh, 22–24 June.

Afsar, R. and Baker, J. 1999. 'Interaction between rural areas and rural towns.' *Background Paper on Prerequisites of Future Swedish Support to Rural Towns Development and Poverty Alleviation*. Norway: AGDER Research Foundation.

Agrawal, A. 2010. 'Local institutions and adaptation to climate change', *Social Dimensions of Climate Change: Equity and Vulnerability in a Warming World*, 2: 173–178.

Ali, I., Hatta, Z.A. and Azman, A. 2014. 'Transforming the local capacity on natural disaster risk reduction in Bangladeshi communities: A social work perspective', *Asian Social Work and Policy Review*, 8(1): 34–42.

Allan, A., Rieu-Clarke, A., Dey, S., Samling, C.L., Ghosh, A., Tagoe, C.A., Nelson, W., Salehin, M. and Mondal, M.S. 2016. 'DECCMA Governance Assessment:

Assessing Governance systems relating to adaptation, migration, climate change and deltas.' DECCMA Working Paper, Deltas, Vulnerability and Climate Change: Migration and Adaptation, IDRC Project Number 107642. Available online at: www.deccma.com. Accessed 28 January 2019.

Alston, M. 2013. 'Women and adaptation', *WIREs Climate Change*, 4(5): 351–358.

Altman, M. and Pannell, K. 2012. 'Policy gaps and theory gaps: Women and migrant domestic labour', *Feminist Economics*, 18(2): 291–315.

Ava, S.K., Uddin, M.S., Rahman, A., Hossain, D., Kumar, S., Mira, S.S. and Haque, M.M.E. 2017. 'Life is Cruel Here. Stories from Forced Migrants in Chittagong, Bangladesh'. Available online at https://kulima.exposure.co/life-is-cruel-here. Accessed 15 November 2018.

Bakewell, O. 2010. 'Some reflections on structure and agency in migration theory', *Journal of Ethnic and Migration Studies*, 36(10): 1689–1708.

Baldwin, A. 2017. 'Climate change, migration, and the crisis of humanism', *WIREs Climate Change*, e460. doi:10.1002/wcc.460.

BBS. 2008. *Statistical Yearbook 2008*. Dhaka: Bangladesh Bureau of Statistics.

Begum, A. 1999. *Destination Dhaka – Urban Migration: Expectations and Reality*. Dhaka: University Press Ltd.

Begum, A. 2015. 'Gender in education: Policy discourse and challenges', *Journal of Development in Practice*, 25(5): 754–768.

Begum, A. 2017. '*Review of migration and resettlement in Bangladesh: Effects of climate change and its impact on gender roles.*' DECCMA Working Paper, Deltas, Vulnerability and Climate Change: Migration and Adaptation, IDRC Project Number 107642. Available online at: www.deccma.com. Accessed 28 January 2019.

Bettini, G. and Gioli, G. 2016. 'Waltz with development: Insights on the developmentalization of climate-induced migration', *Migration and Development*, 5(2): 171–189.

Bhuiyan, M.R.A. and Siddiqui, T. 2015. 'Migration in the Ganges-Brahmaputra-Meghna Delta: A review of the literature.' DECCMA Working Paper, Deltas, Vulnerability and Climate Change: Migration and Adaptation, IDRC Project Number 107642. Available online at: www.deccma.com.

Biao, X. 2007. 'How far are the left-behind left behind? A preliminary study in rural China', *Population, Space and Place*, 13: 179–191.

Black, R., Adger, W.N., Arnell, N.W., Dercon, S., Geddes, A., and Thomas, D. 2011. 'The effect of environmental change on human migration', *Global Environmental Change*, 21: S3–11.

Brown, S., Nicholls, R.J., Lázár, A., Sugata, H., Appeaning, A.K., Hornby, D.D., Hill, C., Haque, A., Caesar, J. and Tompkins, E. 2018. 'What are the implications of sea-level rise for a 1.5°C, 2°C and 3°C rise in global mean temperatures in vulnerable deltas?', *Regional Environmental Change*, 18(6): 1829–1842. doi:10.1007/s10113-018-1311-0.

Bunce, A. and Ford, J. 2015. 'How is adaptation, resilience, and vulnerability research engaging with gender?', *Environmental Research Letters*, 10. doi:10.1088/1748-9326/10/12/123003.

Call, M.A., Gray, C., Yunus, M. and Emch, M. 2017. 'Disruption, not displacement: Environmental variability and temporary migration in Bangladesh', *Global Environmental Change*, 46: 157–165.

Castles, S. 2010. 'Understanding global migration: A social transformation perspective', *Journal of Ethnic and Migration Studies*, 36(10): 1565–1586.

Castles, S. and Miller, M.J. 1993. *The Age of Migration. International Population Movements in the Modern World.* Basingstoke: Macmillan.

Dasgupta, S., Hossain, M.M., Huq, M., and Wheeler, D. 2014. 'Facing the hungry tide: Climate change, livelihood threats, and household responses in coastal Bangladesh.' Policy Research Working Paper 7148, World Bank Group: Development Research Group, Environment and Energy Team. Washington, DC: World Bank.

DSK. 2012. *Current Climate Change Vulnerability: Position of DSK in the Climate Change Discourses in Bangladesh.* Dhaka: DSK. Available online at http://www.dskbangladesh.org/?wpfb_dl=34. Accessed 28 January 2019.

Desai, S. and Banerji, M. 2008. 'Negotiated identities: Male migration and left-behind wives in India', *Journal of Population Research*, 25(3): 337–355.

Etzold, B., Ahmed, A.U., Hassan, S.R. and Neelormi, S. 2014. 'Clouds gather in the sky, but no rain falls. Vulnerability to rainfall variability and food insecurity in Northern Bangladesh and its effects on migration', *Climate and Development*, 6(1): 18–27. doi:10.1080/17565529.2013.833078.

Foresight. 2011. *Migration and Global Environmental Change: Future Challenges and Opportunities.* London: Department of Business Innovation and Skills.

Gemenne, F. 2011. 'Why the numbers don't add up: A review of estimates and predictions of people displaced by environmental changes', *Global Environmental Change*, 21: 41–49.

Gerold-Scheepers, J.F.A. and van Binsbergen, W.M.J. 1978. 'Marxist and non-Marxist approaches to migration in tropical Africa', *African Perspectives*, 21: 35.

GIZ-RMMRU. 2014. 'Climate Change Induced Migration and Urban Informal Settlements.' Policy Briefing Paper No. 8. Published as an output of RMMRU and GIZ Research Climate Migration Study of Resilient and Inclusive Urban Development. Dhaka: RMMRU.

Government of Bangladesh. 2012. *Sixth Five Year Plan of the Government of Bangladesh (2011–2015), Accelerating Growth and Reducing Poverty.* Dhaka: Government of Bangladesh.

Government of Bangladesh. 2015. *Seventh Five Year Plan of the Government of Bangladesh (2016–2020), Accelerating Growth, Empowering Citizens.* Dhaka: Government of Bangladesh.

Gulati, L. 1993. *In the Absence of Their Men: The Impact of Male Migration on Women.* New Dehli: Sage.

Hadi, A. 2001. 'International migration and the change of women's position among the left-behind in rural Bangladesh', *Population, Space and Place*, 7(1): 53–61.

Haque, M.A., Yamamoto, S.S., Malik, A.A. and Sauerborn, R. 2012. 'Households' perception of climate change and human health risks: A community perspective', *Environmental Health* 11(1). doi:10.1186/1476-069X-11-1.

Haug, S. 2008. 'Migration networks and migration decision-making'. *Journal of Ethnic and Migration Studies*, 34(4): 585–605.

Hossain, I.M., Khan, I.A. and Seeley, J. 2003. 'Surviving on Their Feet: Charting the Mobile Livelihoods of the Poor in Rural Bangladesh.' Paper presented at the Conference 'Staying Poor: Chronic Poverty and Development Policy' held at the University of Manchester, 7–9 April.

Hossain, Z., Kazal, M.H. and Ahmed, J.U. 2013. *Rural-Urban Migration and its Implications for Food Security in Bangladesh.* Sylhet: Department of Statistics, Shahjalal University of Science and Technology.

Hugo, G. 2011. 'Future demographic change and its interactions with migration and climate change', *Global Environmental Change*, 21: S21–S33.

Hugo, G.J. 1978. *Population Mobility in West Java*. Yogyakarta: Gadjah Mada University Press.

Hutton, D. and Haque, C.E. 2003. 'Patterns of coping and adaptation among erosion-induced displacees in Bangladesh: Implications for hazard analysis and mitigation', *Natural Hazards*, 29: 405–421.

Islam, M. 2014. 'Women, politics and patriarchy: A case of Bangladesh.' In Alston M. (ed.) *Women, Political Struggles and Gender Equality in South Asia. Gender, Development and Social Change*. London: Palgrave Macmillan.

Joarder, M.A.M. and Miller, P.W. 2013. 'Factors affecting whether environmental migration is temporary or permanent: Evidence from Bangladesh', *Global Environmental Change*, 23 (6): 1511–1524. doi:10.1016/j.gloenvcha.2013.07.026.

Kartiki, K. 2011. 'Climate change and migration: A case study from rural Bangladesh', *Gender & Development*, 19(1): 23–38. doi:10.1080/13552074.20 11.554017.

Laczko, F. and Aghazarm, C. 2009. *Migration, Environment and Climate Change: Assessing the Evidence*. Geneva: IOM.

teLintelo, D.J.H., Gupte, J., McGregor, J.A., Lakshman, R. and Jahan, F. 2018. 'Wellbeing and urban governance: Who fails, survives or thrives in informal settlements in Bangladeshi cities?', *Cities*, 72: 391–402.

Lutz, H. 2010. 'Gender in the migratory process', *Journal of Ethnic and Migration Studies*, 36(10): 1647–1663.

Mahmud, T. and Prowse, M. 2012. 'Corruption in cyclone preparedness and relief efforts in coastal Bangladesh: Lessons for climate adaptation?', *Global Environmental Change*, 22(4): 933–943.

Marshall, R. and Rahman, S. 2012. *Internal Migration in Bangladesh: Character, Drivers and Policy Issues*. Bangladesh: UNDP.

Martin, M., Kang, Y., Billah, M., Siddiqui, T., Black, R. and Kniveton, D. 2013. *Policy analysis: Climate change and migration – Bangladesh: An output of research on climate change related migration in Bangladesh*. Refugee and Migratory Movements Research Unit (RMMRU), University of Dhaka, and Sussex Centre for Migration Research (SCMR), University of Sussex.

Massey, D., Arango, J., Hugo, G., Kouaouci, A., Pellegrino, A. and Taylor, J. 1993. 'Theories of international migration: A review and appraisal', *Population and Development Review*, 19(3): 431–466.

McLeman, R. and Smit, B. 2006. 'Migration as an adaptation to climate change', *Climatic Change*, 76(1–2): 31–53. doi:10.1007/s10584-005-9000-7.

Mia, M.A., Nasrin, S., Zhang, M. and Rasiah, R. 2015. 'Chittagong, Bangladesh', *Cities*, 48: 31–41.

Ministry of Expatriates' Welfare and Overseas Employment; International Organization for Migration (IOM). 2004. *Institutionalizing Diaspora linkage: The emigrant Bangladeshis in UK and USA*. Dhaka: IOM.

Myers, N. 2005. 'Environmental refugees: An emergent issue', *13th Economic Forum*, 1–5.

Pearse, R. 2017. 'Gender and climate change', *WIREs Climate Change*, 8: e451. doi:10.1002/wcc.451.

Perch-Nielsen, S.L., Bättig, M.B. and Imboden, D. 2008. 'Exploring the link between climate change and migration', *Climatic Change*, 91: 375. doi:10.1007/s10584-008-9416-y.

Rahman, H.Z., Hossain, M. and Sen, B. 1996. *Dynamics of Rural Poverty in Bangladesh, Dhaka*. Dhaka: Bangladesh Institute of Development Studies (BIDS).

Rao, N. 2014. 'Migration, mobility and changing power relations: aspirations and praxis of Bangladeshi migrants', *Gender, Place & Culture*, 21(7): 872–887.

Rao, N. and Mitra, A. 2013. 'Migration, representations and social relations: Experiences of Jharkhand labour to western Uttar Pradesh', *Journal of Development Studies*, 49(6): 37–41.

Ruyssen, I. and Salomone, S. 2018. 'Female migration: A way out of discrimination?', *Journal of Development Economics*, 130: 224–241. doi:10.1016/j.jdeveco.2017.10.010.

Saul, B. 2012. 'The Security Risks of Climate Change Displacement in Bangladesh.' Sydney Law School Research Paper No. 12/58. Available at SSRN: https://ssrn.com/abstract=2138006. Accessed 28 January 2019.

Shamsuddoha, M., Hossain, S. and Shahjahan, M. 2013. 'Study Report on Land Availability for Climate Displaced Communities of Bangladesh.' Young Power in Social Action.

Sharma, M. and Zaman, H. 2009. '*Who migrates overseas and is it worth their while? An assessment of household survey data from Bangladesh.*' Policy Research Working Paper 5018. Policy Reduction and Economic Management Network, Poverty Reduction Group. Washington, DC: World Bank.

Siddiqui, T. 2001. *Transcending Boundaries: Labour Migration of Women from Bangladesh*. Dhaka: University Press Limited.

Siddiqui, T. 2003. *Migration as a Livelihood Strategy of the Poor: The Bangladesh Case*. Dhaka: RMMRU.

Siddiqui, T. and Mahmud, R.A. 2014. *Impact of Migration on Poverty and Local Development in Bangladesh*. Dhaka: RMMRU.

Stark, O. and Bloom, D.E. 1985. 'The new economics of labour migration', *The American Economic Review*, 75(2): Papers and *Proceedings of the Ninety-Seventh Annual Meeting of the American Economic Association*, 173–178.

Stern Review Team. 2006. *What Is the Economics of Climate Change?* London: HM Treasury.

Stojanov, R., Kelman, I., Ahsan Ullah, A.K.M., Dusi, B., Procházka, D. and Kavanová Blahutová, K. 2016. 'Local expert perceptions of migration as a climate change adaptation in Bangladesh'. *Sustainability*, 8: 1223. doi:10.3390/su8121223.

Stott, C. and Huq, S. 2014. 'Knowledge flows in climate change adaptation: Exploring friction between scales', *Climate and Development*, 6(4): 382–387. doi:10.1080/17565529.2014.951014.

Szabo, S., Renaud, F.G., Hossain, S., Sebesvari, Z., Matthews, Z., Foufoula-Georgiou, E. and Nicholls, R.J. 2015. 'Sustainable development goals offer new opportunities for tropical delta regions'. *Environmental Science and Policy*, 57: 16–23. doi:10.1080/00139157.2015.1048142.

Szabo, S., Nicholls, R.J., Sebesvari, Z., Brondizio, E., Matthews, Z., Foufoula-Georgiou, E., Renaud, F.G., Tessler, Z., da Costa, S., Hetrick, S., Tejedor, A. and Dearing, J.A. 2016. 'Population dynamics, delta vulnerability and environmental change: comparison of the Mekong, Ganges–Brahmaputra and Amazon delta regions', *Sustainability Science*, doi:10.1007/s11625-016-0372-6.

Szabo, S., Adger, W.N. and Matthews, Z. 2018. 'Home is where the money goes: Migration-related urban-rural integration in delta regions', *Migration and Development*, doi:10.1080/21632324.2017.1374506.

Tacoli, C. 2009. 'Crisis or adaptation? Migration and climate change in a context of high mobility', *Environment and Urbanization*, 21(2): 513–525.

Todaro, M.P. 1969. 'A model of labour migration and urban unemployment in less developed countries', *American Economic Review*, 59(1): 138–148.

Todaro, M.P. 1976. *Internal Migration in Developing Countries*. Geneva: ILO.

Van Aelst, K. and Holvoet, N. 2016. 'Intersections of gender and marital status in accessing climate change adaptation: Evidence from rural Tanzania', *World Development*, 79: 40–50. doi:10.1016/j.worlddev.2015.11.003.

Warner, K. and Afifi, T. 2014. 'Where the rain falls: Evidence from 8 countries on how vulnerable households use migration to manage the risk of rainfall variability and food insecurity', *Climate and Development*, 6(1): 1–17. doi:10.1080/17565529.2013.835707.

Williams, N. 2009. 'Education, gender, and migration in the context of social change', *Social Science Research*, 38(4): 883–896. doi:10.1016/j.ssresearch.2009.04.005.

Willis, K. and Yeoh, B. 2000. *Gender and Migration*. Cheltenham/Camberley: Edward Elgar Publishing.

World Bank. 2010. *The Economics of Adaptation to Climate Change*. Washington, DC: World Bank.

Yeoh, B.S.A. and Ramdas, K. 2014. 'Gender, migration, mobility and transnationalism', *Gender, Place & Culture*, 21(10): 1197–1213.

9 Women-headed households, migration and adaptation to climate change in the Mahanadi Delta, India[1]

Sugata Hazra, Amrita Patel, Shouvik Das, Asha Hans, Amit Ghosh, and Jasmine Giri

Introduction

Given the socially determined roles of men and women, contemporary gendered research on the social impacts of climate change enables a better understanding of their differential vulnerabilities, coping mechanisms and adaptation processes. This chapter focuses on women-headed households in relation to vulnerability, migration and adaptation in the context of climate change in the Mahanadi Delta (MD) in India, which comprises five coastal districts in the state of Odisha: Bhadrak, Kendrapara, Jagatsinghpur, Puri and Khordha.

Recent research has contributed to analyzing the impact of climate variability in the identified geographical locations where gendered household differentials are observed as a consequence of their adaptive capacities and sensitivity (IPCC 2014; Kuppannan et al. 2015; Solomon and Rao 2018). A systemic disadvantage is reflected in both policy and research where women's vulnerability is seen to pose obstacles to their wellbeing and very survival in climate change situations (Alhassan et al. 2019; Morchain et al. 2015; Singh 2019). In this situation the biggest challenge is gendered loss of livelihood and food security forcing people to migrate (Afifi et al. 2016; Duncan et al. 2017; Rao and Mitra 2013).

Investigation of households in migrant situations from a gendered perspective been limited, and though it is increasing – albeit with a variety

1 The authors thank DECCMA-India team members who provided insight and expertise that was immensely helpful for the research.
This work is carried out under the Deltas, Vulnerability and Climate Change: Migration and Adaptation (DECCMA) project (IDRC 107642) under the Collaborative Adaptation Research Initiative in Africa and Asia (CARIAA) programme with financial support from the UK Government's Department for international Development (DfID) and the International Development Research Centre (IDRC), Canada. The views expressed in this work are those of the creators and do not necessarily represent those of DfID and IDRC or its Board of Governors.

of conflicting assumptions and conclusions – the overall understanding at inter- and intra-household levels is limited to a few authors (Desai and Banerji 2008; Pottier 1994). It is pertinent to recognize that the issue of headship is often addressed with a lack of clear methodology. In most of the research on gender and climate change, women and men are seen in binary terms and treated as undifferentiated homogeneous entities. Several authors highlight the multidimensionality of gender relationships and how they are shaped by various intersectionalities, such as of caste, class, social status or age, while others have investigated gender disparity within households, where women-headed households are found to be more vulnerable due to gaps in income and in access to social and economic resources (Mitra 2018; Rao and Hans 2018; Williams 2015). Recently a new line of thought considering male vulnerability and poverty as important areas for investigation has emerged (Chant 2003; Chant 2015; Nabikolo et al. 2012; Oginni et al. 2013). These different positions nevertheless illustrate that gender disparities and disadvantages specific to women-headed households combine to make them economically vulnerable in general, and more so in the context of climate shocks.

The limitation in existing research on migration and climate change is that while impact on men- and women-headed households is examined, there exists no analysis of women-headed households with and without migrants as a separate category, based on an extensive database. The other omission has been in the context of adaptation, where there has been no analytical space for the success of adaptation strategies. In this analyses, the suggested gender of the head of the household is not merely a reference point as it does have significant socio-economic impact implications (Gangopadhyay and Wadhwa 2004). First, women-headed households, as illustrated in existing literature, may have more dependents and thus have lower worker/non-worker ratios than other households (Klasen et al. 2015). Second, due to gender biases, women heads might have to work for low wages and have less access to assets and productive resources, such as land, financial capital and technology, than men, threatening the wellbeing of the household (Agarwal 1994; Alhassan et al. 2019). Third, women typically bear the burden of household chores and look after children and dependents without other support, resulting in time and mobility constraints, unlike in men-headed households (Buvinić and Gupta 1997; Rosenhouse 1989). Overall, women-headed households seem poorer than men-headed households (Gangopadhyay and Wadhwa 2004; Unisa and Datta 2005).

Data remain an important indicator of the gendered complexity from different levels. The Census of India categorizes the head of household by marital status: 'never married', 'currently married', 'widowed', 'divorced' and 'separated'. The definition of the head of household for census purposes is a person who is recognized as such by the household.[1] According to the Census, the number of women-headed households in the country is increasing. Between 2001 and 2011, women-headed households rose from 10.4 per cent to 13.2 per cent of total households. In Odisha, the proportion of women-headed households also increased from 10.1 per cent to 12.5 per cent of total households during the same period.[2] Among the women household heads, widows comprise 68.3 per cent and 64.5 per cent nationally and in the state of Odisha, respectively. In the delta districts, the proportion of women-headed households increased from 9.4 per cent to 12 per cent between 2001 and 2011, registering a growth rate of 6 per cent annually (from 0.13 million in 2001 to 0.21 million in 2011). Widows in the five delta districts average 57.6 per cent and 56.6 per cent respectively in the 2001 and 2011 Censuses. The National Family Health Survey (NFHS) data also show an increase in women-headed households:[3] 9.2 per cent, 10.3 per cent and 14.4 per cent in NFHS I (1992–93), II (1998–99) and III (2005–06) respectively (Kumar and Gupta 2012).

A majority (78.8 per cent) of the 23 million women-headed households[4] (12.8 per cent of total households) surveyed in rural India in the Socio-Economic Caste Census (SECC), had a monthly income less than Rs. 5,000 (Government of India SECC 2011a, b). Overall, 14 million women-headed households are 'deprived' according to the SECC, where deprivation criteria are based on conditions of housing, landlessness, absence of an able-bodied adult member, any adult male member or a literate adult (Maqsoodi 2015). Poverty is affected by gender and the choice of poverty measures does not influence the results (Julka and Das 2015). However, despite the poverty (Nagla 2008), it is argued that women in women-headed households successfully network to gain information and support, which suggests higher confidence and decision-making powers, factors that play an important role in adaptation processes. In the climate-change policy framework of India, the National Action Plan (NAPCC) recognizes that the impacts of climate change could prove particularly severe for women. Women-headed households, however, are not mentioned and all women are projected as a homogeneous category, undifferentiated by caste, class, age, marital status or region (Government of India 2008).[5]

This chapter begins with an introduction on climate change and women-headed households. The second section provides the administrative, demographic and socio-economic characteristics of the study area. The third section establishes the link between environmental risk and migration. The fourth section describes the data and the methodological approach of the household survey. The fifth section discusses the research findings on vulnerability, migration and adaptation activities of the women-headed households, and the final section concludes by highlighting the chapter's contributions to the literature as well as the implications of the findings for policy making.

The Mahanadi Delta

The Mahanadi Delta (MD) comprises a network of three major rivers – the Mahanadi, the Brahmani and the Baitarini – draining into the Bay of Bengal, covering a coastline of 200 km, and stretching from the south near Chilika, the largest coastal lagoon in Asia, up to the Dhamra river in the north. The study area under the Deltas, Vulnerability, Climate Change: Migration and Adaptation (DECCMA)[6] project in the MD consists of 45 sub-districts (community development blocks) within the five coastal districts of Bhadrak, Kendrapara, Jagatsinghpur, Puri and Khordha (Figure 9.1), which is delineated based on the 5m contour line and administrative boundaries (Lazar et al. 2015). With an area of 13,000 km², it covers 8 per cent of Odisha's geographical area.

In 2011 the population of MD was 8 million, a population density much higher than that of Odisha (270 persons per km²) and India (382 persons per km²). Almost 78 per cent of the population is rural. They are mainly dependent on agriculture, working as cultivators and agricultural labourers.

As far as the situation of women is concerned, their work participation rate is low in general, which is important to note for the purpose of adaptation strategies and policy intervention. The gender gap is stark, as the male work participation rate of the five districts is high (55.4 per cent), whereas women's participation is only 12.3 per cent, much lower than the state average of 27.2 per cent (Government of India 2011a). The larger proportion of women workers are engaged in household industries both in rural and

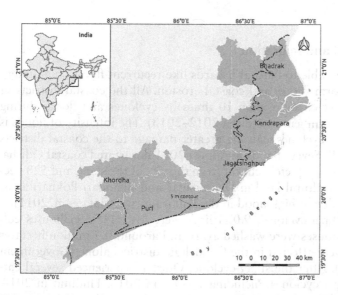

Figure 9.1 The study area map of Mahanadi Delta.

urban areas. Employment generated in favour of women under Mahatma Gandhi National Rural Employment Guarantee Scheme (MGNREGS) in the MD is low (12.22lakh man-days) in comparison to districts with high tribal population like Mayurbhanj (22.16 lakh man-days). The number of self-help groups functioning with government support in delta districts is also low, being less than half that of non-delta districts like Ganjam and Mayurbhanj (Government of Odisha, 2014–15). This picture of women's low work participation in all fields and low productive engagement contrasts with women's literacy,[7] which is high in the MD (about 80 per cent) (IIPS 2015–16).

Furthermore, the average proportion of women-headed households over the five delta districts is 12.5 per cent, but Kendrapara district has a high percentage of women-headed households (15.3 per cent), of whom 46.4 per cent are currently married. While the district of Puri has a high percentage (64 per cent) of widow-headed households, the four other delta districts have a lower percentage (55 per cent on average) (Government of India 2011a). Thus the status of women in the delta districts presents a complex picture with high literacy rates and low work participation of women, accompanied by the presence of a significant proportion of widows and married women heads in some districts. Currently married women-headed households indicate the absence of men who are most likely to be long-term migrants. This scenario merits a nuanced understanding not only of the gender dynamics but also of the heterogeneity of the women, specifically the women heads of households. This chapter seeks to analyze the social realities in the context of climate change through household survey results (Table 9.1).

Coastal hazard and migration

The MD is vulnerable to natural hazards like recurrent floods, high-intensity cyclones, storm surges and coastal erosion. All the coastal districts of Odisha were affected more than 10 times by cyclones and floods during 1995–2012 (Government of Odisha 2012–2013). The intensity of floods is severe during the cyclones, causing greater damage to the coastal districts (Bahinipati 2014). Severe floods in August 2011 devastated coastal Odisha. 877 villages were completely cut off from the rest of the state, and 528 hectares (ha) of agricultural land in Dhamnagar and Bhandari Pokhari areas were inundated due to high flood in the Baitarani river (Jata et al. 2011).

In the1999 'super-cyclone', 9,078 lives were lost, 445,595 houses collapsed, 13,762 houses were washed away and around 0.7 million hectares of agriculture were affected in the five deltaic districts alone (Government of Orissa 2004). After the super-cyclone, Odisha experienced several catastrophic tropical cyclones, including Phailin in 2013, Hudhud in 2014, Titli in 2018 and Fani in 2019. These cyclones adversely impacted people's livelihoods. Women, being at the centre of household management, were the worst sufferers (Iwasaki 2016; Patra et al. 2013; Swain et al. 2016).

Table 9.1 Socio-economic profile of the five districts of the Mahanadi Delta

	Odisha	Bhadrak	Jagatsinghpur	Kendrapara	Khurdha	Puri	Average of 5 districts	Source
Female literacy rate (for population 7 yrs and above)	64	75.8	80.6	79	81.6	78.3	79.1	Census 2011
Work participation Rate — Male	56.1	53.9	56.3	53.8	55.3	57.6	55.4	Census 2011
Work participation Rate — Female	27.2	7.9	14	11.1	13.5	14.8	12.3	Census 2011
Physical achievement under NRLM in 2013-14 — Num of SHG functioning	271559	8404	10124	9922	11539	10273		Economic Survey Odisha, 2014–15
Employment generation under MGNEGS in 2013-14 - employment generated in lakh mandays — Women	238.9	3.1	1.9	1.7	1.9	1.6		Economic Survey Odisha, 2014–15
Women headed HH as % of Total HH	12.5	11.5	12.2	15.3	11.2	10.3	12.1	Census 2011
Currently married Women headed HHs as % of Total Female headed HHs	26	34.3	35.5	46.4	34.1	28.5		Census 2011
Widowed women headed households		57.4	57.2	47.8	56.7	64		Census 2011

The main livelihood of this delta is agriculture and the majority of farmers are smallholders, who are exposed to frequent climate hazards and socio-economic stresses (Duncan et al. 2017).

Both fast-onset climatic hazards like cyclone, surge or flood and slow-onset hazards like sea-level rise, coastal land loss, heatwaves and droughts potentially displace large numbers of people from the coast globally (Doocy et al. 2013; Needhams et al. 2015; Rahman et al. 2015). Odisha is no exception. Migration, mostly internal, can be permanent or temporary. It is often difficult to assess the exact number of out-migrants from census data. Second, it is a difficult task to identify whether the out-migration is any way linked to propensity and risk of climatic hazards and variability.

In the present research under the DECCMA project, following the Intergovernmental Panel on Climate Change (IPCC) fifth assessment report (AR 5), risk and net migration have been analyzed using the indirect method and mapped at the sub-district level in MD to understand the link between migration and climate risk. According to IPCC AR 5, risk is the function of three factors: hazard (H), vulnerability (V) and exposure (Ex) (IPCC 2014). The term vulnerability has been split into 'sensitivity' and 'adaptive capacity' for the simplification of the conceptual and methodological frameworks of risk. Flood, cyclone and coastal erosion which are the common environmental stressors in MD are included in the hazard category of the risk assessment. Population density, agricultural dependency and road density are used as exposure, sensitivity and adaptive capacity, respectively, for the risk assessment. Net migration (the difference between in-migration and out-migration) has been estimated using the vital statistics method (indirect method), in the absence of data on place of birth, place of enumeration and age/sex-disaggregated data at the sub-district level. From the Geographical Information System (GIS) overlay analysis, a close correspondence of high-risk zones with negative net migration (more out-migration or sending areas) can be observed, while the low-risk zones in the delta mostly appear as net receiving areas with more in-migration (Figure 9.2).

Several sub-districts where out-migration dominates, such as Dhamnagar, Ersama, Balikuda and Tihidi, are bio-physically and socio-economically at very high risk (Das et al. 2016). Seasonal migration from these coastal districts results from the repeated crop loss and lack of returns from existing livelihoods (Das et al. 2016). While male migration dominates the trend, women's migration is picking up gradually from vulnerable parts of MD. Many women who have school education are now going to Puri or Bhubaneswar for work in small industries (Das et al. 2016; Samling et al. 2015). Movement of young girls, particularly those who are unmarried and have some skills, to garment industries in Bengaluru, Tiruppur and Chennai is also seen.[8]

Khordha, which is the most urbanized district in Odisha with 42.9 per cent urban population, Puri, a famous destination for religious tourism and Paradip, a growing sea port, have emerged as the preferred destination areas for migrants coming from adjoining rural communities of MD

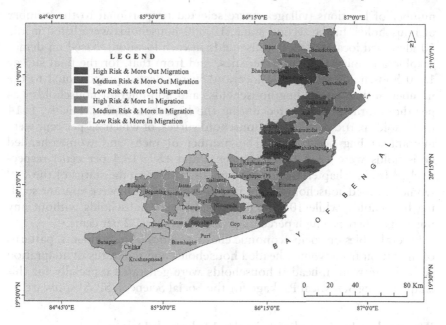

Figure 9.2 Multi hazard risk and migration map of Mahanadi Delta.

for greater economic opportunities. Migrants have also moved to Chennai, Gujarat, Tamil Nadu and Kerala (inter-state) and Qatar, Saudi Arabia and Dubai (international). Climate change can thus already be seen to be affecting people living in the region; the migration dynamics, specifically the out-migration of males and the emergence of women heads of households, are visible. Women living in rural areas are more dependent on natural resources and agriculture for their livelihood. It is therefore necessary to understand the complexity of links between the geo-physical, agro-ecological and socio-economic situations with regard to women, particularly for women left behind due to male migration, if workable policies for them are to be suggested.

Taking the lead from the secondary data analysis, an extensive household survey in the five coastal districts of MD was undertaken during 2016 for better understanding of the ground realities of migration, climatic hazard perception, vulnerabilities and scope of in-situ adaptation with a gendered perspective.

Household survey: methodology and data analysis

Fifty locations in the five Districts of the MD were identified by applying a two-stage cluster sampling technique for the implementation of the household survey. The first stage was a village-level multi-hazard map (flood, cyclone, erosion), which divided the MD (study area) into five multi-hazard zones (very low, low, medium, high and very high). In each of the zones, a number of locations (villages) were selected proportional to the number of households. In the second stage, 10,000 households were listed in the 50 identified locations (200 households in each location), based on demographic and migration characteristics, and from that, for the final survey 1500 households (30 in each location) were selected proportional to the number of migrant-sending households[9] and non-migrant households[10] as per the information collected during the household listing. Finally, 1414 households participated in the household survey, of which 22 per cent were migrant-sending households. The number of men- and women-headed households were 1225 (86.6 per cent) and 189 (13.4 per cent) respectively,[11] In this chapter, we restrict our discussions to the status of the 189 women-headed households, 82 (43.3 per cent) of whom were migrant-sending households, while 107 (56.6 per cent) were households without any migrants. There is a high percentage of widows (60.32 per cent).

Several tables on socio-economic characteristics of respondents, patterns of migration from women-headed households and the status of adaptation activities in women-headed households were generated especially for this chapter using Statistical Package for the Social Sciences (SPSS) software.

Women-headed households in the Mahanadi Delta

The distribution of respondents for men- and women-headed households according to socio-economic characteristics is presented in Table 9.2. The social characteristics include household size, age, marital status, caste and formal education.

Among social characteristics, household size plays a significant role as it influences the income and expenditure of a household. Survey data show that average household size is 4.87 for men-headed households and 4.27 for women-headed households, and a smaller 3.46 for widow-headed households. Age is another important variable as it has a direct relationship with social and environmental awareness. Almost 32 per cent of the women heading a household are below the age of 40. For both men and women heads of households, the highest percentage is found between the ages of 40 and 60 years; around 25 per cent of heads of households are over 60 years. Widows' vulnerability is revealed through age, with those aged 40 and above constituting 89 per cent.

Marital status plays a gendered role in the society. According to the survey of the 1414 households, more than 90 per cent of male household heads were currently married, but only 32.8 per cent of the women heads were, given the general restrictions on the remarriage of widows (60.3 per cent) (Chen 1998). Education is another significant variable associated with better economic and psychological outcomes. In keeping with the national trend of census data, the education level of male respondents was observed to be much higher than that of the female. Women respondents from women-headed households with no education (41.3 per cent) are greater in number than those with primary and secondary education in the sample. More

Table 9.2 Socio-economic characteristics of respondents (N =1414)

	Characteristics	Men Headed Households (%)	Women Headed Households (%)	Widow Headed Households (%)
Household Size	1–3	22.06	38.62	55.26
	4–6	63.73	51.32	37.72
	≥ 7	14.22	10.05	7.02
	Mean Household Size	4.87	4.27	3.46
Age (Years)	≤ 40	26.2	31.75	13.16
	41–60	47.84	38.62	44.74
	> 60	25.88	29.63	42.11
	Mean Age	51.95	51.48	58.15
Marital Status	*Never married*	1.8	1.59	–
	Currently married	93.47	32.8	–
	Widowed	4.57	60.32	100.00
	Divorced/ Abandoned / separated	0.08	5.29	–
Caste	*General*	34.23	33.33	33.33
	Scheduled Caste (SC)	24.35	20.63	21.93
	Scheduled Tribe (ST)	1.8	2.65	3.51
	Other backward caste (OBC)	39.62	43.39	41.23
Formal Education	*No Education*	11.59	41.27	54.39
	Primary Education	38.53	30.16	32.46
	Secondary Education	36.16	23.81	10.53
	Higher Education	13.63	4.76	2.63
Main Livelihood	*Farmer & Fishermen*	36.08	5.29	6.14
	Regular Salaried Employee	8.57	5.29	3.51
	Small Business Owner	7.84	4.76	3.51
	Construction & Factory Workers	21.55	4.76	2.63
	Unpaid Home Carer	0.82	36.51	20.18
	Others (Retired, Pension holder, Supported by Relatives/Family, Unemployed, etc.)	25.06	43.39	64.04
Monthly Income (Indian Rupees)	*No Income*	3.43	36.51	18.42
	≤ 3000	22.06	46.56	62.28
	3001–6000	40.36	12.7	15.79
	> 6000	34.15	4.23	3.51
	Mean Monthly Income	6077.68	1364.95	1487.86

(Continued)

Table 9.2 (Continued)

Characteristics		Men Headed Households (%)	Women Headed Households (%)	Widow Headed Households (%)
House Area (Square Metres)	< 50	37.29	44.97	39.47
	50–100	37.04	31.75	32.46
	> 100	25.76	23.28	28.07
	Own House	96	93.12	92.98
Farm Size (Hectares)	0	28.92	31.22	78.07
	0.01–1.00	58.58	60.85	14.91
	1.00+	12.5	7.94	7.02
Total Observations		1225 (100.0)	189 (100.0)	114 (100.0)

Source: Primary survey, 2017.

than half the widows were illiterate and 32.46 per cent had studied up to the end of primary education.

In the present study, the economic characteristics of the households included main livelihood, monthly income, house area, farm size and ownership. Low literacy and income are linked, hence a huge disparity in monthly income is observed between men- and women-headed households. Men respondents were mostly farmers and fishermen (36.1 per cent) and construction and factory workers (21.6 per cent), whereas women (including widow-headed household respondents) were mainly unpaid home carers, given their household care tasks and restrictions on mobility. The mean monthly income of the men heads is Rs. 6078, whereas it is only Rs. 1365 for women heads (1 Indian rupee = 0.014US$, Sept. 2019). More widow-headed households (62 per cent) had income which was lower (earning less than Rs. 3000) than women-headed households (47 per cent). In contrast, not even 1 per cent of men stayed at home as unpaid carers, and only 3.4 per cent were without income.

The majority of the respondents do own houses, but small ones (less than 50m² in area). Almost 30 per cent of total respondents have no farm land. More widow-headed households managed to own their home than women-headed households, which could be due to government housing schemes for poor women. Most did not, however, have their own farm land, and in the absence of an alternate livelihood option in the region, male migration is common.

Vulnerability and women-headed households

Recent research in climate change has indicated that vulnerabilities are gendered and women are typically portrayed as vulnerable, rather than as negotiating and dealing regularly with different kinds of changes in their lives (Okali and Naess 2013; Rao et al. 2019). A vulnerability assessment helps us to know whether women-headed households are more vulnerable than their men-headed counterparts. Both bio-physical and socio-economic vulnerabilities have been assessed in this study. Bio-physical vulnerability is a function of the frequency and severity of climatic hazards, while social vulnerability is the inherent property of a system arising from its internal characteristics (Brooks et al. 2005). 28 theoretically important and policy-relevant bio-physical and socio-economic variables (sub-components) were selected under five major components – climate variability, environmental stressors, impact of extreme events, socio-demographic profile and economic status. The bio-physical variables are perceptions of respondents, whereas the socio-economic variables are responses that are tangible and measurable.[12] All the responses are in percentages and have a positive (+) functional relationship with major components of vulnerability, which means the higher the value, the higher the vulnerability. A description of the variables used in the present study is provided in Table 9.3.

In the bio-physical vulnerability index (BPVI) assessment survey, respondents stated that there had been a change in precipitation and an increase in temperature over the previous five years, but the perception was slightly higher in the women-headed households than in the men-headed ones. More than 50 per cent of respondents (all men-, women- and widow-headed households) reported experiences of environmental stressors (flood, cyclone and erosion) with increasing frequency and impact in the last 10 years. These extreme events directly caused livelihood shocks. The survey data indicate that the women- and widow-headed households reported more monetary losses as well as loss of life than men-headed households in extreme events, due to crop failure, and livestock and equipment damage (Table 9.2).

Similarly, for the socio-economic vulnerability index (SEVI) assessment, survey data reveal that most of the women-headed households are socially vulnerable, and the percentage of young and elderly illiterate members is also higher than in the men-headed households. Elderly people and children are dependents, have mobility constraints and lack resilience. Lower education or illiteracy constrains their ability to understand warning information and access to recovery information. The presence of widows/divorcees or separated women is substantially higher in the women-headed households. All these factors make women-headed households more sensitive and susceptible to extreme environmental events. It is observed that most of the women heads have no income or savings, and work as unpaid home carers. The existence of a large number of non-workers in both men- and women-headed households contributes to a slower recovery from any hazard event, whether of fast or slow onset. Lower resilience or adaptive capacity is often characterised by unsafe drinking water, poor sanitation facilities and the predominance of "kutcha" houses (thatched houses with roofs made of hay/leaves/branches/jute bags/plastic or other materials) – conditions which are typical for women-headed households. Food insecurity is a high risk; eating one meal a day was

Table 9.3 Vulnerability Assessment of Men and Women Headed Households in MD

Variables	Men Headed Households (N = 1225)	Women Headed Households (N = 189)	Women Headed Households	
			Widow Headed Households (N = 114)	Non-Widow Headed Households (N = 75)
	%	%	%	%
Bio-Physical Vulnerability Index (BPVI)				
Climate Variability (last 5 years)				
Increase in temperature	98.53	99.47	99.12	100.00
Change in precipitation	97.22	96.83	95.61	98.67
Average	97.88	98.15	97.37	99.33
Environmental Stressors (last 10 years)				
Flood	78.76	80.95	75.44	89.33
Drought	69.28	64.02	60.71	70.67
Erosion	30.8	26.46	20.35	36.99
Salinity	12.99	18.52	17.86	20.27
Storm surges	60.13	53.44	54.46	54.05
Cyclone	91.18	91.01	89.38	94.67
Average	57.19	55.73	53.03	61.00
Impact - Damages (Monetary Loss)/Loss of life				
Crops	70.18	70.37	68.15	75.93
Livestock	21.98	25.4	26.67	22.22
Equipment and other material assets	52.94	58.73	57.04	48.15
Loss of life	4.98	9.52	14.04	2.67
Average	37.52	41.01	41.47	37.24
BPVI	57.41	57.89	56.57	59.47

Socio-Economic Vulnerability Index (SEVI)					
Socio-Demographic Profile					
Household size (≥ 5)	35.19	40.74	39.15	50.65
Female members (> 50%)	31.48	54.07	47.62	27.7
Aged person (60+ years) (>20%)	16.67	27.41	24.34	21.49
Children (< 15 years) (> 30%)	42.59	30.37	33.86	31.78
Widow/Divorced/Separated (> 10%)			69.31	15.11
Illiteracy (>30%)	25.93	42.22	37.57	26.06
Socially backward class (SC, ST, OBC)	70.37	65.19	66.67	65.77
Chronically ill-injured-disabled	75.93	72.59	73.02	74.75
Average	42.59	47.51	48.94	39.16
Economic Status				
Agricultural dependency (≥ 50%)	14.81	11.85	12.17	38.4
Household head without income	70.37	22.96	36.51	3.51
Non-workers (> 50%)	33.33	36.30	35.45	34.07
Food insecurity (eaten 1 meal per day or less)	44.44	42.96	43.39	42.81
Roof (other than cement, brick, asbestos)	48.15	45.19	40.74	30.07
Unsafe drinking water (pond, canal, spring, etc.)	3.70	8.89	7.41	7.92
No sanitation facility	68.52	63.70	64.02	62.66
No support from NGO or Govt.	70.37	69.63	69.84	72.71
Average	44.21	37.69	38.69	36.52
SEVI	43.46	42.27	43.82	37.84
Composite Vulnerability Index (CVI)	51.46	49.42	50.86	47.63

Source: Primary survey, 2017.

reported more by the adult women in women-headed households than in widow-headed ones, who had fewer household members. There is high pressure of work on women who are carers in all three types of household due to a high incidence of chronic illness or disability in the family. The survey reported that over the last five years, almost 70 per cent of households had not received any assistance from the government or NGOs for improvement of their home (wall, roof and floor) or livelihood diversification. The insecurity resulting from poor access to food is heightened for women-headed households by a combination of malnutrition, inadequate sanitation and the inability to pay for health services due to low or no income.

The chapter has developed a composite vulnerability index (CVI) which combines the bio-physical and socio-economic vulnerability indices. Results show that the BPVI values are similar for men- and women-headed households, which indicates that they are bio-physically in the same vulnerable situation. The SEVI value of women-headed households (43.8 per cent) is much higher than that of men-headed households (37.8 per cent). It can be said that women-headed households are socio-economically in more vulnerable situations than men-headed households in MD. The value of the CVI as derived in the present chapter is greater for the women-headed households (50.9 per cent) than the men-headed households (47.6 per cent) and widow-headed households (49.4 per cent). The lower CVI of widowed households can be explained by their lower dependency ratio, alongside lower dependence on agriculture (Figure 9.3). The ability of women-headed households to absorb losses and be resilient to environmental impacts is very low. Due to their low work participation and income, and significant family care responsibilities, women heads have a more difficult time with recovery than men. The data show that overall, widow-headed households are vulnerable, but sustain themselves as they get support, while women-headed households without support or pensions remain highly vulnerable.

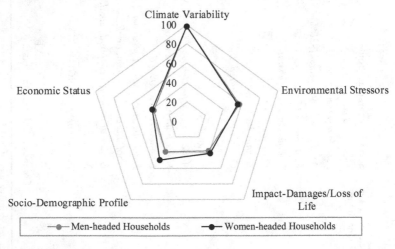

Figure 9.3 Major components in vulnerability assessment.

Migration and women-headed households

Migration is one of the options which human populations use to adapt to environmental changes (McLeman and Hunter 2010). It can potentially reduce the socio-economic and bio-physical vulnerabilities of women-headed households. As stated earlier, among the 1414 households who participated in the household survey, 22 per cent have sent migrants, and the percentage of migrant-sending households was high in the 'very high' and 'high' multi-hazard zones of the delta (Figure 9.4), indicating that people are moving from vulnerable parts of the MD to live and work in safer places.

Out of the total 189 women-headed households, 82 (43.4 per cent) are with migrants and 107 (56.6 per cent) are without migrants. A significant number of women household heads have acquired their present status because of the out-migration of men, making it imperative to discuss whether migration reduces or aggravates the vulnerability of women-headed households.

Seasonal migration (56.58 per cent) dominates the migration pattern among women-headed households (Table 9.4). From women-headed households, men (74.85 per cent) mostly migrate once or twice a year leaving the women behind to take on the additional responsibilities of dependents and households. Women migrants from women-headed households are 25.15 per cent, mostly students, unpaid home carers or unemployed.

Both men and women migrants with secondary education and in the age group 21–40 migrate more from women-headed households. In case of women migrants, the decision to migrate depends on the household. They mostly move with their family. It is the elderly people, children and women who stay behind as the 'trapped population' in the delta. With diminishing returns from traditional agriculture and fisheries in the face of climatic

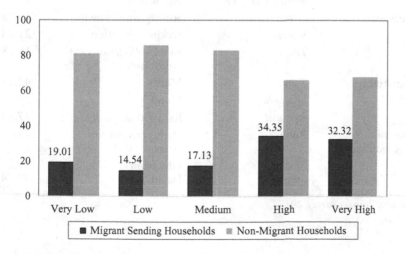

Figure 9.4 Distribution of households in five multi-hazard zones.

Table 9.4 Patterns of Migration from Women Headed Households in MD (*N* = 189)

Responses			%
Type		Seasonal migration	56.58
Frequency		1-2 times	45.9
Duration		3-6 months	37.1
Scale		Internal Migration	98.68
Destinations	*State*	Odisha, West Bengal, Karnataka, Delhi	69.74
	District	Khordha, Kolkata, Puri	65.12
	City	Bhubaneswar, Puri (M), Jagatsinghpur (M)	64.81
Current Migrant's Characteristics	Male Migrants		74.85
	Age	21–40	54.92
	Marital Status	Currently Married	59.02
	Education	Secondary	47.41
	Livelihood	Factory Worker, Regular Salaried Employee, Construction Worker	77.88
	Monthly Income	Rs. 10000 and Below	70.87
	Female Migrants		25.15
	Age	21–40	53.66
	Marital Status	Currently Married	58.54
	Education	Secondary	44.74
	Livelihood	Unpaid Home Carer, Student, Unemployed	100
	Monthly Income	No Income	100
Reasons	*First*	Seeking employment	41.78
	Second	Seeking education	21.23
	Third	Family obligations / problems	19.18
Remittances	*Type*	Money	42.11
	Frequency	Monthly	50
	Amount	Rs. 5000 and Below	37.84
Uses of Remittances	*Rank-1*	Daily consumption (food, bills)	73.68
	Rank-2	Education	39.29
	Rank-3	Health care	40.74

Source: Primary survey, 2017.

uncertainties, migration is emerging as an alternative livelihood strategy in the MD, according to the respondents.

The immediate reason behind migration is economic (41.78 per cent), with the majority of the men migrating in search of employment opportunities. Friends and family members who have already migrated temporarily provide an information network for migration. The second most frequently mentioned reason is education, with 21 per cent of respondents reporting that the migrant left to study for a degree or to obtain training in a new skill. The survey results also provide evidence of female migration (25.51 per cent) from the delta, though mostly as unaccompanied spouses and home carers.

However, migration often shows a dual and contradictory impact on the vulnerability of the women-headed households. While women left behind due to the out-migration of men constitute a vulnerable group by itself, monthly remittances sent by migrant male members have the potential to alleviate their vulnerable status, at least marginally. The household survey revealed that remittances improve standards of living by enabling them to pay for food, education and health care. Consequently, thanks to remittances, the monthly per-capita expenditure (MPCE) of women-headed households with migrants was higher (Rs. 2355) than that of those without any migrant members (Rs. 1473). Most of the household survey respondents mentioned that migration improves the social and economic status of migrant-sending households. In other words, among women-headed households, those with migrants, mostly men, are better off than those without. More than 85 per cent of total respondents felt migration is helpful for migrants also. All the respondents in the household survey reported that migration improves economic security, education and work opportunities for migrants, and it also brings new ideas and practices to the village when migrants return. Thus, the exchange of money, knowledge and ideas between the migrant's places of origin and destination can potentially reduce the socio-economic and bio-physical vulnerabilities of the women-headed households. Migration, then, has both positive and negative impacts on human development (Figure 9.5a and 9.5b).

While there are several positive outcomes of migration in the women-headed households and the community at large, the respondents felt that it left very few young men in the village, reducing the community's ability to face slow- or fast-onset climatic hazards and adversities. With more than 25 per cent of the surveyed households expressing their intention to migrate in the future, the propensity to migration in the villages of MD will increase in future, leaving more women-headed households behind to cope with and adapt in situ to climate change and climatic disasters. The social vulnerability of women-headed households, both with and without migrants, requires consideration and has implications for policy support.

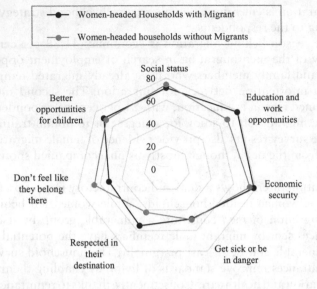

Figure 9.5a Dimensions of migration at individual level (migrant).

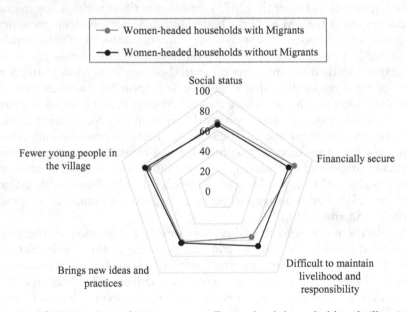

Figure 9.5b Dimensions of migration at collective level (household and village).

Adaptation and women-headed households

Adaptation experiences are gendered and even though exposure to climate variations may be the same for men and women in any given location, there are gender-based differences in vulnerability and consequently in adaptation and adaptive capacity (Adger et al. 2005; Muttarak et al. 2015). It has been observed, for instance, that men and women farmers in many developing countries have different levels of vulnerability and adaptive capacity to climate change (Denton 2002). In the fifth IPCC assessment report, adaptation is described as the "process of adjustment to actual or expected climate and its effects" (IPCC 2014:39). Though not gendered, it is apparent that adaptation, in IPCC terms, is an intervention to adjust to climate change effects.

At the local level, the Odisha 2010–15 Climate Change Action Plan (OCCAP) recognized the impact of climate change on women in the context of water scarcity and reducing availability of biomass, but did not address the issue of adaptation (Government of Odisha 2010–2015). In relation to the key issues of adaptation to coastal disaster, water stress, declining agriculture and health, women's perspectives were hardly discussed. The adaptation activities included in the action plan are neither drawn from women's needs nor specify any. The survey sought the views of the women respondents with regard to 21 adaptation activities, of which agriculture-related adaptation activities were asked only for relevant households (Table 9.5). The recall period was the previous five years (2011–16).

The assumption that migration is positive would mean that women-headed household gain from migration. Adopting this intersectional approach, the women-headed households with and without migrants were asked in the survey about adaptation activities and whether they were better off after taking up the said activity. Success was calculated based on whether the household had improved by taking up the particular adaptation activity. The activities shown in Table 9.5 were related to increase in livelihood, capacity building and building infrastructure.

Table 9.5 shows that both types of households – women-headed with and without migrants – succeeded equally in some activities, such as taking loans, using hired labour and higher amounts of fertilizer, going outside to work, sending a household member to work outside and receiving assistance from NGOs and government. With future safety as a strategy in this disaster-prone area in mind, both planted trees around the house, though more women without migrants did so, moved house and stayed with the community. New gender roles were also noticed in households without migrants, as women have taken up male-oriented activities such as hiring labour and carrying out irrigation work.

Households with migrants were highly successful in crop diversification and using irrigation. They adapted and in an attempt to make the situation better made use of insurance and joined cooperatives. These were related to their livelihood and were enabled as in all the households, migrant members returned. Together both types of household successfully used adaptation activities such as crop diversification, climate-tolerant crops, tree plantations around the home, protection against natural hazards (own organized/community shelter). These activities improved household finances, education, health and access to government or NGO assistance.

Table 9.5 Adaptation Activities of Women Headed Households in the last 5 years

Adaptation Activities	Women Headed Households with Migrants (N = 82)		Women Headed Households without Migrants (N = 107)		Total (N = 189)	
	Yes (%)	Success (%)	Yes (%)	Success (%)	Yes (%)	Success (%)
1 Loan to purchase things for the household, or to improve livelihood	42.68	54.29	27.1	58.62	33.86	56.25
2 Insurance for main livelihood	17.07	92.86	12.15	53.85	14.29	74.07
3 Joined a **cooperative** in relation to main livelihood	20.73	76.47	16.82	66.67	18.52	71.43
4 Modified, or improved **House** (walls, roof, floor)	39.02	65.63	34.58	75.68	36.51	71.01
5 **Cut down trees** around home	13.41	54.55	12.15	53.85	12.7	54.17
6 **Planted trees** around home	25.61	95.24	33.64	72.22	30.16	80.7
7 Using hired **labour** to support in generating income	15.85	69.23	9.35	70	12.17	69.57
8 **Women** working outside the village	3.66	66.67	14.95	68.75	10.05	68.42
9 **Moved to a new house** within the same village	3.66	100	3.74	75	3.7	85.71
10 Sent a household member to **work outside the village**	58.54	68.75	14.02	66.67	33.33	68.25
11 Household member **come back permanently**	25	100	0	0	12.5	100

(Continued)

Table 9.5 (Continued)

Adaptation Activities	Women Headed Households with Migrants (N = 82)		Women Headed Households without Migrants (N = 107)		Total (N = 189)	
	Yes (%)	Success (%)	Yes (%)	Success (%)	Yes (%)	Success (%)
12 Crop diversification	16.67	100	0	0	5.56	100
13 Planting climate tolerant crops	0	0	8.33	100	5.56	100
14 Increased used of fertiliser	50	66.67	66.67	62.5	61.11	63.64
15 Put in irrigation	33.33	100	33.33	50	33.33	66.67
16 Mixed farming/ fishing production	22.22	0	0	0	8.7	0
17 Bought farming / fishing equipment	0	0	0	0	0	0
18 Training on new fishing or farming skills/ methods	0	0	0	0	0	0
19 Fished new breeds / used new breeds in ponds	0	0	0	0	0	0
20 Received government or NGO assistance	25.61	80.95	24.3	84.62	24.87	82.98
21 Organised own protection or used a community shelter	25.61	61.9	26.17	85.71	25.93	75.51
Average	20.89	59.68	16.06	49.72	18.23	61.35

Source: Primary survey, 2017.

The average success percentage (all adaptation activities) for women-headed households with migrants was 59.67, higher than for those without migrants, which indicates that migration has the potential to reduce the overall vulnerability of women-headed households by enhancing their adaptive capacity and agency to address climate variability. This view is strengthened by their strong advocacy of working outside the village for both men and women.

Activities which require more time, money and labour are mostly adopted by women-headed households with migrants. The perception that women are agency-less is eliminated, as an important gendered role adoption is visible. This is an attempt at breaking barriers, though it must be recognized that women's mobility remains restricted and patriarchy remains in place (Figure 9.4), with women migrating from this region not for economic reasons but as carers of families.

Conclusion: women-headed households in climate change

While it is widely accepted that climate change may promote migration from coastal areas in the world as well as in India, this study was able to establish a relationship between high risk of climate hazards and higher migration in the deltaic region of Odisha. The primary survey of 1414 households also confirmed a similar trend in the temporary or short-term migration of men, resulting in an increasing number of women-headed households. This gender-disaggregated data set, the first of its kind for the MD, has given rise to certain observations which have significant implications for future research as well as policy interventions. The bio-physical and socio-economic vulnerability indices, as well as the composite vulnerability index, give an in-depth understanding of the vulnerability of women-headed households in the context of climatic changes. The status of widows and their vulnerabilities and strengths have also been analyzed in the chapter.

An important point learnt from the study is that both women- and widow-headed households are more vulnerable to the impacts of climate change in the delta than households headed by men, mainly for socio-economic reasons. However, an important conclusion that the study draws from a household perspective is the lower vulnerability of widow-headed households than women-headed households, as the latter have less support and no steady income.

The risks faced by women- and widow-headed households due to their household responsibilities as unpaid carers and their inability to access income keep them poor. With non-existent policy initiatives, –no farm skill development in the sending areas, limited access to land and no insurance cover, they remain deprived of their economic rights. Feminization of poverty in women-headed households is not a new phenomenon, but it has not been a national or international target for eradication, even when women remain food insecure and unpaid carers, as seen in this study.

Second, in spite of their high vulnerability for socio-economic reasons, it has been highlighted that women-headed households do take on male roles when it comes to adaptation activities. Though it is a positive attribute, we have to take into account the increase in pressure of work on these women. Since they are the main carers of children, larger numbers of women in the household and chronically ill or disabled members of the family, and they live in a patriarchal system where men rarely do domestic work, the gender bias in which masculine privilege prevails continues in post-climatic hazard situations. This can give them less time to socially network which they need not only for a better quality of life but also to access socially derived resources. It may also restrict women's personal savings which can be used in emergencies or for personal/children's needs. These are markers of vulnerability that are hidden in research or in the strategies adopted for overcoming them.

Women-headed households remain highly vulnerable in climate change situations as the impact of adaptation has not improved their overall situation. Successful adaptations identified include crop diversification and farming practices, but in-situ adaptation has not succeeded, resulting in migration. Women-headed households have thus supported the idea of migration, but restricted mobility has been a barrier that has not been overcome until recently, when younger women are found to be migrating. It can be concluded from the study that women-headed households with migrants, if managed and monitored properly, have the potential to alleviate their vulnerability. Remittances received regularly or intermittently by both women and women-headed households partially meet their everyday needs and reduce their economic vulnerability to some extent. However, as there is no effective policy for internal migrants in the country, these remittances remain largely uncertain for the women-headed households left behind.

In conclusion, the important idea that the chapter draws from a household perspective on the feminization of poverty is the change for extremely deprived widow-headed households who, with economic support, increase their ability to rise from the low level on which society places them. Here, social change due to the transformation of gender roles also signifies a more gender-equitable role in the context of women-headed households in general, as they do take on male roles in adaptation which, though not economically viable, does entail social change. It cannot also be overlooked that men may also be affected, due to their inability to pay for health services and clean sanitation, especially in migrant destinations. This element of powerlessness and stress for men needs further exploration.

Notes

1 Head of the household: She or he is generally the person who bears the chief responsibility for managing the affairs of the household and takes decision on behalf of the household. The head of household need not necessarily be the oldest male member or an earning member, but may be a female or a younger member of either sex. In case of an absentee *de jure* 'Head' who is not eligible to be enumerated in the household, the person on whom the responsibility of managing the affairs of household rests was to be regarded as the head irrespective whether the person is male or female (Government of India data. Key words).
2 Calculated from Table HH-06 Households By Marital Status, Sex and Age of the Head of Household, Household Series, Census of India, 2011a.
3 National Family Health Survey (NFHS) publishes data on headship based on self-reported survey.
4 According to the Government of India census SECC 2011, if any female member of a household desires or declares herself as a separate household, she is treated as a separate household (Govt of India SECC 2011b).

5 India's National Action Plan on Climate Change (NAPCC) states, "the impacts of climate change could prove particularly severe for women. With climate change, there would be increasing scarcity of water, reductions in yields of forest biomass, and increased risks to human health with children, women and the elderly in a household becoming the most vulnerable, special attention should be paid to the aspects of gender" (Government of India, 2008: 14)

6 DECCMA project explores whether migration is an adaptation option in such low-lying deltaic regions and aims to provide policy support to create conditions for sustainable gender sensitive adaptation (DECCMA Brief, 2017, www.deccma.com). Study sites include the Ganges-Brahmaputra-Meghna Delta in Bangladesh and India, Mahanadi Delta in India and Volta Delta in Ghana (www.cariaa.net).

7 The state female literacy rate is 64.01 per cent as per the 2011 Census (Government of India, 2011a).

8 The participants in the stakeholder workshops in Bhadrak and Puri districts under the DECCMA project work in the MD made this observation.

9 A migrant-sending household is any household that has members that have migrated from the sending area in the last 10 years. This includes current migrants and returned migrants as per the DECCMA's household survey methodology.

10 Non-migrant-sending households are those which have no members who have migrated from the sending area in the last 10 years as per the DECCMA's household survey methodology.

11 The following definitions are used in the DCCMA household survey methodology: Women-Headed Household: Women-headed households are those in which an adult female has the most authority and responsibility for household affairs or earns the most income in the household. Male-Headed Household: Male-headed households are those in which an adult male has the most authority and responsibility for household affairs or earns the most income in the household. If the household head has migrated away, then the head is the person who has the most influence over household affairs in their absence, or who earns the most income.

12 The indices on bio-physical and socio-economic vulnerability are based on indicators by which the respondents provided their perception in the survey questionnaire.

References

Adger, W.N., N. Brooks, G. Bentham, M. Agnew and S. Eriksen. 2005. *New Indicators of Vulnerability and Adaptive Capacity*. Norwich: Tyndall Centre for Climate Change Research.

Afifi, T., A. Milan, B. Etzold, B. Schraven, C. Rademacher-Schulz, P. Sakdapolrak, A. Reif, K. Geest and K. Warner. 2016. 'Human Mobility in Response to Rainfall Variability: Opportunities for Migration as a Successful Adaptation Strategy in Eight Case Studies', *Migration and Development*, 5(2): 254–274. doi:10.1080/21 632324.2015.1022974

Agarwal, B. 1994. *A Field of One's Own: Gender and Land Rights in South Asia*. Cambridge South Asian Studies. Cambridge: Cambridge University Press.

Alhassan, S., J. Kuwornu and Y. Osei-Asare. 2019. 'Gender Dimension of Vulnerability to Climate Change and Variability: Empirical Evidence of Smallholder Farming Households in Ghana', *International Journal of Climate Change Strategies and Management*, 11(2): 195–214. https://doi.org/10.1108/IJCCSM-10-2016-0156. Accessed 20 October 2019.

Arora-Jonson, S. 2011. 'Virtue and Vulnerability: Discourses on Women, Gender and Climate Change', *Global Environmental Change*, 21: 744–751.

Brooks, N., W.N. Adger and P.M. Kelly. 2005. 'The Determinants of Vulnerability and Adaptive Capacity at the National Level and the Implications for Adaptation', *Global Environmental Change*, 15(2): 151–163.

Buvinić, M. and G.R. Gupta. 1997. 'Female-headed Households and Female-Maintained Families: Are They Worth Targeting to Reduce Poverty in Developing Countries?', *Economic Development and Cultural Change*, 45(2): 259–280.

Chant, S. 2003. *'Female Household Headship and the Feminisation of Poverty: Facts, Fictions and Forward Strategies.'* New Working Paper Series (9). Gender Institute, London School of Economics and Political Science, London, UK.

Chant, S. 2015. 'Female Headed Households Leadership as an Asset? Interrogating the Intersections of Urbanization, Gender, and Domestic Transformation', in Caroline Masor (ed.) *Gender, Asset Accumulation and Just Cities.* Routledge, pp. 33–51.

Chen, M.A. (ed.) 1998. *Widows in India: Social Neglect and Public Action.* New Delhi: Sage.

Das, S., S. Hazra, T. Ghosh, S. Hazra and A. Ghosh. 2016. *'Migration as an Adaptation to Climate Change in Mahanadi Delta',* Abstract Number: ABSSUB-98. *Paper presented at the Conference 'Adaptation Future 2016',* Rotterdam, Netherlands, 10–13 May.

Denton, F. 2002. 'Climate Change Vulnerability, Impacts, and Adaptation: Why Does Gender Matter?', *Gender and Development*, 10(2): 10–20.

Desai, S. and M. Banerji. 2008. 'Negotiated Identities: Male Migration and Left-Behind Behind Wives in India', *Journal of Population Research*, 25: 490–499.

Doocy, S., A. Daniels, S. Murray and T.D. Kirsch. 2013. 'The Human Impact of Floods: A Historical Review of Events 1980–2009 and Systematic Literature Review', *PLOS Currents Disaster*, 16 April. https://pdfs.semanticscholar.org/1251/3c8b56b6434c-b1eeea46e0cfa9539ac085e1.pdf?_ga=2.95288660.93114147.1560941461-1709343239.1560941461. Accessed 19 June 2019.

Duncan, J., E. Tompkins, J. Dash and B. Tripathy. 2017. 'Resilience to Hazards: Rice Farmers in the Mahanadi Delta, India.' *Ecology and Society*, 22(4): 3. https://doi.org/10.5751/ES-09559-220403. Accessed 10 September 2019.

Gangopadhyay, S. and W. Wadhwa. 2004. *Are Indian Female-headed Households more Vulnerable to Poverty?* New Delhi: Bazaar Chintan.

Government of India. 2008. *National Action Plan on Climate Change*, Prime Minister's Council on Climate Change, Ministry of Environment, Forests and Climate Change. http://www.moef.nic.in/modules/about-the-ministry/CCD/NAP_E.pdf. Accessed 22 June 2017.

Government of India. 2011a. *Census of India.* Office of the Registrar General and Census Commissioner, Government of India.

Government of India. 2011b. *Census of India.* Office of the Registrar General and Census Commissioner, Government of India. http://www.censusindia.gov.in/2011census/Hlo-series/HH06.html. Accessed 2 September 2018.

Government of India. SECC 2011a. *Socio Economic and Caste Census (SECC).* New Delhi: Ministry of Rural Development, Government of India. http://secc.gov.in/categorywiseIncomeSlabReport?reportType=Female percent20Headed percent20Category. Accessed 2 September 2018.

Government of India. SECC 2011b. *Socio Economic and Caste Census (SECC).* New Delhi: Ministry of Rural Development, Government of India. http://secc.gov.in/faqReportlist. Accessed 1 September 2018.

Government of India data. Key words. https://data.gov.in/keywords/female-headed-household. Accessed 1 September 2018.

Government of Odisha. 2013. *Annual Report on Natural Calamities (2012–2013).* Bhubaneswar: Special Relief Commissioner, Revenue and Disaster Management Department. http://srcodisha.nic.in/annualReport/fuxOK3oyAnnual_Report_2012-13%20on%20NC.pdf. Accessed 10 September 2019.

Government of Odisha. 2014–2015. *Odisha Economic Survey.* Bhubaneswar: Planning & Coordination Department, Directorate of Economics and Statistics. http://www.indiaenvironmentportal.org.in/files/file/Odisha%20Economic_Survey_2014-15.pdf. Accessed 10 September 2019.

Government of Odisha. 2010–2015. *Climate Change Action Plan (OCCAP).* Bhubaneswar: Department of Forest and Environment. https://forest.odisha.gov.in/ActionPlan/CCAP%20ORISSA%20FINAL-1.pdf. Accessed 10 September 2018.

Government of Orissa. 2004. *Human Development Report.* Bhubaneswar: Planning and Coordination Department. http://devfocus.in/wp-content/uploads/2014/12/human_development_report_2004_orissa_full_report.pdf. Accessed 10 September 2019.

IIPS. 2015–16. *National Family Health Survey (NFHS-4), State Fact Sheet Odisha.* Mumbai: International Institute for Population Sciences (IIPS). http://rchiips.org/NFHS/pdf/NFHS4/OR_FactSheet.pdf. Accessed 8 August 2018.

IPCC. 2014. 'Climate Change 2014: Impacts, Adaptation, and Vulnerability. Part A: Global and Sectoral Aspects.' Contribution of Working Group II to Field, C.B., V.R. Barros, D.J. Dokken, K.J. Mach, M.D. Mastrandrea, T.E. Bilir, M. Chatterjee, K.L. Ebi, Y.O. Estrada, R.C. Genova, B. Girma, E.S. Kissel, A.N. Levy, S. MacCracken, P.R. Mastrandrea, and L.L. White (eds) *Fifth Assessment Report of the Intergovernmental Panel on Climate Change.* Cambridge/New York: Cambridge University Press, pp. 1132–1197.

Iwasaki, S. 2016. 'Linking Disaster Management to Livelihood Security against Tropical Cyclones: A Case Study on Odisha State in India', *International Journal of Disaster Risk Reduction,* 19: 57–63.

Jata, S.K., M. Nedunchezhiyan and N. Jata. 2011. 'Flood and Risk Management Strategy in Orissa', *Orissa Review,* September–October: 62–66.

Julka P. and S. Das. 2015. '*Female Headed Households and Poverty: Analysis Using Household Level Data*'. Working paper 133. Chennai: Madras Institute of Development Studies (MIDS).

Klasen, S., T. Lechtenfeld and F. Povel. 2015. 'A feminization of Vulnerability? Female headship, poverty, and vulnerability in Thailand and Vietnam', *World Development,* 71: 36–53.

Kumar, N. and A. Gupta. 2012. '*Female Headed Household in India: Evidence Based Situational Analysis.' Population Association of America Annual Meeting Program,* San Francisco.

Kuppannan, P., A. Haileslassie and K. Kakumanu. 2015. 'Climate Change, Gender and Adaptation Strategies in Dryland Systems of South Asia: A Household Level Analysis in Andhra Pradesh, Karnataka and Rajasthan States of India.' Research Report No. 65 ICRISAT Research Program Resilient Dryland System. Retrieved from https://cgspace.cgiar.org/handle/10568/71194, 11 October 2019.

Lazar, A.N., R.J. Nicholls, A. Payo et al. 2015. 'A Method to Assess Migration and Adaptation in Deltas: Preliminary Fast Track Assessment' (No. 107642). DECCMA working paper, Deltas, Vulnerability and Climate Change: Migration and Adaptation, IDRC project.

Maqsoodi, A. 2015. 'Census Reveals Gloomy Picture of Life in Female-Headed Households' 6 July 2015. http://www.livemint.com/Politics/RjAdjOgWkNMqHGI1DqX8tJ/Census-reveals-gloomy-picture-of-life-in-femaleheaded-house.html. Accessed 1 September 2018.

McLeman, R.A. and L.M. Hunter. 2010 'Migration in the Context of Vulnerability and Adaptation to Climate Change Insights from Analogues', *Wiley Interdisciplinary Reviews: Climate Change*, 1(13): 450–461.

Mitra, A. 2018. 'Male Migrants and Women Farmers in Gorakhpur', *Economic and Political Weekly*, 53(17): 55–62.

Morchain, D. G. Prati, F. Kelsey and L. Ravon 2015. 'What if Gender Became an Essential, Standard Element of Vulnerability Assessments?', *Gender and Development*, 23(3): 481–496.

Muttarak, R., W. Lutz and L. Jiang. 2015. 'What Can Demographers Contribute to the Study of Vulnerability?', *Vienna Yearbook of Population Research*, 13: 1–13.

Nabikolo, D., B. Bashaasha, M. Mangheni and J.G.M. Majaliwa. 2012. 'Determinants of Climate Change Adaptation among Male and Female Headed Farm Households in Eastern Uganda', *African Crop Science Journal*, 20(2): 203–212.

Nagla, M. 2008. 'Male Migration and Emerging Female Headed Families', *Asian Women*, 24(1): 1–23.

Needham, H.F., B.D. Keimand and D. Sathiaraj. 2015. 'A Review of Tropical Cyclone-Generated Storm Surges: Global Data Sources, Observations, and Impacts', *Reviews of Geophysics*, 53(2): 545–591.

Oginni, A., B. Ahonsi, F. Ukwuije. 2013. 'Are Female-Headed Households Typically Poorer than Male-Headed Households in Nigeria?', *Journal of Socio-Economics*, 45: 132–137.

Okali, C. and L.O. Naess. 2013. 'Making Sense of Gender, Climate Change and Agriculture in Sub-Saharan Africa: Creating Gender Responsive Climate Adaptation Policy.' Future Agricultures Working Paper 57.

Parikh, J.K., D.K. Upadhyay, T. Singh. 2012. 'Gender Perspectives on Climate Change and Human Security in India: An Analysis of National Missions on Change', *Cadmus*, 1(4): 180–186.

Patra, M., Tripathy, S., and Jena, I. 2013. 'Health hazards by sea cyclones in Odisha, the supercyclone and the Phailin', *Odisha Review*, 70(4): 30–37.

Pottier, J. 1994. '*Understanding food stress at local levels*.' In *Food Systems under Stress in Africa. African Canadian Research Cooperation; proceedings of a workshop held in Ottawa, ON, Canada, 7–8 Nov. 1993*. Ottawa, ON, Canada: IDRC.

Rahman, M.K., B.K. Paul, A. Curtis and T.W. Schmidlin. 2015. 'Linking Coastal Disasters and Migration: A Case Study of Kutubdia Island, Bangladesh', *The Professional Geographer*, 67(2): 218–228.

Rao, N. and A. Hans. 2018. 'Gender and Climate Change: Emergent Issues for Research, Policy and Practice', *Economic and Political Weekly*, 53(17): 35–37.

Rao, N. and A. Mitra. 2013. 'Migration, Representations and Social Relationships: Experiences in Jharkhand Labour to Western Uttar Pradesh', *Journal of Development Studies*, 49(6): 846–860.

Rao, N., E.T. Lawson, W.N. Raditloaneng, D. Solomon and M.N. Angula. 2019. 'Gendered Vulnerabilities to Climate Change: Insights from the Semi-Arid Regions of Africa and Asia', *Climate and Development*, 11(1): 1426. doi: 10.1080/17565529.2017.1372266.

Rosenhouse, S. 1989. '*Identifying the Poor: Is Headship a Useful Concept? Living Standards Measurement Study*.' Working Paper 58. Washington, DC: The World Bank.

Samling, C.L., S. Das and S. Hazra. 2015. 'Migration in the Indian Bengal Delta and the Mahanadi Delta: a review of the literature.' DECCMA Working Paper, Deltas, Vulnerability and Climate Change: Migration and Adaptation, IDRC Project Number 107642. Available online at: www.deccma.com

Singh, C. 2019. 'Migration as a Driver of Changing Household Structures: Implications for Local Livelihoods and Adaptation', *Migration and Development*, 8(3): 301–319.

Solomon, D. and N. Rao. 2018. 'Wells and Wellbeing: Gender Dimensions of Groundwater Dependence in South India', *Economic and Political Weekly*, 53: 38–45.

Swain, U., M. Swain and Sahoo, R. H. 2016.'Experiences of Women with Super Cyclone in CoastalOdisha.Women and Disasters in South Asia', *Survival, Security and Development*, 99: 111–154.

Unisa, S. and N. Datta. 2005. Female Headship in India: Levels, Differentials and Impact. *25th Conference of IUSSP in July*. France. http://www.demoscope.ru/weekly/knigi/tours_2005/papers/iussp2005s50705.pdf. Accessed 15 September 2019.

Williams, M. 2015. *Gender and Climate Change Financing: Coming out of the Margin*. London/New York: Routledge.

Annexure 9

To get a single value of vulnerability from all the above mentioned bio-physical and socio-economic variables, three methodological steps of vulnerability assessment have been followed in this chapter.

First step of the assessment to get the value for each major component using the concept of *simple arithmetic mean*:

$$AI = \sum S_i / N \tag{A.1}$$

Where, AI is the value of average index (score) for each major component, Si represents the sub-components (i denotes the ith sub-components) and N is the total number of sub-components in each major component.

In second step, the bio-physical and socio-economic vulnerability indexes are determined by using the concept of *weighted arithmetic mean*:

$$VI = \sum (W_i.AI_i) / \sum W_i \tag{A.2}$$

Where, VI is Vulnerability Index (both bio-physical and socio-economic), AI_i is the major component indexed by i, Wi is the weight of each major component (number of sub-components).

In final step, Composite Vulnerability Index (CVI) which is the combination of bio-physical and socio-economic vulnerability indexes has been calculated using *simple arithmetic mean* (Eq. A.1).

10 Gender dynamics and climate variability

Mapping the linkages in the Upper Ganga Basin in Uttarakhand, India

Vani Rijhwani, Divya Sharma, Neha Khandekar, Roshan Rathod, and Mini Govindan[1]

Introduction

It is well established that climate change has differential impacts on people. Related disasters and impacts often intensify existing inequalities, vulnerabilities, economic poverty and unequal power relations (Brody et al. 2008; IPCC 2014). Differently positioned women and men perceive and experience climate change in diverse ways because of their distinct socially constructed gender roles, responsibilities, status and identities, which often results in varied impacts, coping strategies and responses (Lambrou and Laub 2004). There is also an understanding that lack of political influence, access to resources, gender-based labour roles and embedded cultural norms combine to render women and girls particularly vulnerable to extreme weather and other climate-related events (Adger 1999; Björnberg and Hansson 2013).

The Hindu Kush Himalayas (HKH), an area known as a climate change 'hot spot', provides irrigation services with the monsoon resulting in 75–90% of the water consumed for major cereal supplies in the world (Singh et al. 2011). Over the last 100 years, Himalayan warming has surpassed the global average of 0.74 ° C (IPCC 2007). This combined with regional variability (Immerzeel et al. 2010; Pellicciotti et al. 2010), geographical heterogeneity and immense socio-cultural diversity could adversely affect the livelihoods of mountain communities, especially because they are highly dependent on natural resources for their livelihood and their very survival (Neumayer and Plümper 2007; Ahmed and Fajber 2009; Arora-Jonsson 2011). Women are the most affected due to their centrality in these communities and the gendered nature of their household responsibilities (Aguilar 2009; Ogra and Badola 2015).

Hydro-meteorological information in the HKH is extremely limited and the microclimatic variations noted at elevation suggest that they are not representative of local climatic conditions (Singh et al. 2011). There are few hydro-meteorological stations, especially at higher elevations, and inadequate long-term weather data. Simulation studies and forecasts are of limited use to the community in terms of shaping responses to future changes. After studies in Eastern Tibetan communities, Byg and Salick (2009) considered that awareness of local attitudes was essential to gaining a better understanding of the impact of climate change and recognizing problems previously ignored as a consequence of confusion about the extent and effect of weather variables at a particular location and the intensity of climate impact.

Communities whose lives and livelihoods rely on eco-system services, especially farmers, are conscious of changes and have accumulated local knowledge over decades (Chaudhary and Bawa 2011; Vedwan and Rhoades 2001), as shown in other studies conducted in the Himalayas or other mountainous regions of the world (Orlove et al. 2010). Such knowledge can contribute to empirical climate modelling and lead to increasing the coarse spatial resolution of climate predictions, important for researchers and policy makers.

Our study is built on the assumptions that (a) climate variability and change is already noticeable to, and directly affecting the livelihoods of, the people in the study area; (b) climate variability is not a new phenomenon to mountain people and they have developed a wealth of traditional response strategies; (c) change in the Himalayas is driven by a variety of environmental and non-environmental factors, not only climate; and (d) vulnerability to climate change is unevenly distributed across and within communities. Based on these assumptions, we examine gendered institutions to discuss differences between men's and women's lived experiences as formed by various social realities. This necessitates an understanding of gendered institutions and conceptualizing gender as a central part of the social framework (Scott 1986) and an integral part of community perception and processes.

Gender is, like all institutions, a result of people having different roles and conflicting interests and identities. Framing gender as a social institution thus shows how change is both resisted and accomplished over time and draws attention to the multiple features – ideology, practices, constraints, conflicts and power – affirming its complexities and multifaceted nature. Nonetheless, this change is guided by the perception of the population. Framing gender as a social institution promotes an exploration of the collective nature of institutions and people and discourages their separation into macro and micro domains. Further, gendered institutions is a notion used to explain how gender relations and the construction of femininity and masculinity are entrenched in daily institutional processes and practices (Lowndes and Roberts 2013). North (1990) defines institutions as "the rules of the game"; hence, gendered institutions are rules, norms and practices that affect, among others, the behaviour, activities, roles and relations of men and women, respectively, in differentiated ways.

In this chapter, we attempt to present an approach that uses gendered institutions to answer three questions in the context of the Uttarakhand, Himalayas: i) what are the caste-specific and gendered differentiated livelihood attributes in case study villages at specific elevations? ii) What are the prominent action situations in that area/elevation? What are the 'rules in use' pertaining to actors belonging to specific castes and genders? iii) what are the perceived changes in climate and its impacts on these attributes and 'rules in use', and what are the patterns emerging from the analysis?

Framework and methodology

Overview of the Institutional Analysis and Development (IAD) framework – a gendered approach

We utilize the above disaggregated conception of gendered institutions, rules, norms and effects to identify the corresponding interplay by applying the Institutional Analysis and Development (IAD) framework (Ostrom 1990). The framework strengthens existing analyses on gender roles and responsibilities, furthering understanding of how gendered dynamics shape differential vulnerability in the context of perceived climate change in rural agrarian settings across three elevations of the Garhwal region of Uttarakhand. Our categorization of elements within each study site according to the IAD framework yields the analytical framework in Figure 10.1.

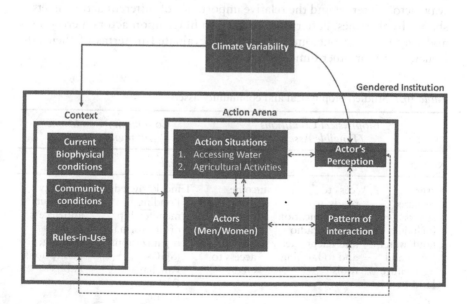

Figure 10.1 Conceptual framework of the study.

Source: modified and adopted from Ostrom et al., 1994, p. 1.

The conceptual unit of the IAD framework is called the **Action Arena**. The action arena is shaped by an **action situation** that involves **actors** and their **interactions** in a particular context following certain rules. The action situation is referred to as a social space where actors interact, exchange goods and services, solve problems or fight; the actors are those who participate in the situation. In the present context, gendered institutions represent the action arena; action situations are activities related to agriculture and accessing water for domestic purposes; and the actors are men and women in the study sites.

The action arena is conceptualized amid a context of external factors representing the initial conditions that actors face (as presented in the first box of Figure 10.1). It establishes the possible pattern of interactions and actors' perceptions that are yielded within a particular situation. The initial conditions are broadly categorized by the underlying **biophysical conditions, community conditions** and **rules in use** that shape the access to and activities the actors perform. Biophysical conditions focus on tangible assets (e.g., natural, physical, and financial assets) both in terms of location and seasonal distribution present within the study site as presented in Table 10.1. Community conditions are intangible assets such as time, knowledge, social standing, networks, perception and habits (ways of carrying oneself and interacting with others) (Di Gregorio et al. 2019). These are broadly categorized into social and human assets. Rules in use refer to a range of prevailing social norms (rules) within a particular action situation and decision-making arrangements, both formal and informal. The ways actors interact and the relative importance of different action assets is shaped by the rules. Patterns of interaction hinge upon actors' perceptions and may result in outcomes that can be evaluated in terms of their efficiency, equity or other dimensions.

Table 10.1 Studied biophysical and community assets

Biophysical Conditions (Tangible Assets)			Community Conditions (Intangible Assets)	
Natural	*Physical*	*Financial*	*Social*	*Human*
Natural resource stocks including land, water, forest and agriculture land etc.	Access to roads, to communication such as radio and telephone, and to farming tools	Income (from on- and non-farm activities), access to credit	Time, knowledge Bonding, membership of (in) formal organizations (SHGs, Panchayat Raj) and linking networks within the village	The ability to labour and access to skills training, perceptions, habits

Moreover, gendered institutions directly or indirectly shape and are shaped by the interactions between actors and action situations. Our study framework presents the key patterns emerging from interactions within the gendered institutions in all our study sites.

Further, the IAD framework also conceptualizes multiple levels of analysis by considering nested characteristics of rules governing the institutional arrangements. It postulates three levels of institutional rules: operational, collective choice and constitutional choice. Operational rules define decisions corresponding to the set of questions on how, when, what and where. Collective choice rules define decisions that correspond to how operational rules can be changed and who can change them. Constitutional choice rules govern the eligibility of participants to change and define the collective choice rules. These rules help us inform the level of interaction between different action settings, day-to-day decision-making processes and the associated climate effect. For the present study, only the operational rules are considered. Actions are taken or decisions about future actions are made by individual actors operating at this level and are based on the set of institutional arrangements within which they operate, such as the intra-household division of labour, as explained in subsequent sections of this chapter. Who can change the rules and who is eligible to change the rules is beyond the purview of the current study, as it requires deeper studies over several points in time of the traditional and cultural system of the gendered institutions.

Data collection and analysis

The study employed a range of data collection techniques to build a narrative on perceived change in climatic and weather parameters and to understand the current roles and responsibilities and access regimes associated with both men and women across communities. These techniques include participatory rural appraisal (PRA) tools, semi-structured household interviews, focus group discussions and key informant interviews. Secondary data were obtained from various sources like the 2011 national census and the 2011 socio-economic caste census (Census of India 2011 and SECC 2011).

Transect walks were undertaken in the villages to build rapport with the people. Information on the distribution of household settlements of different ethnic groups (Table 10.2) and the different biophysical and community assets presented in Table 10.1 (natural, physical, financial, human and social) in the landscape were recorded during these walks. Further, wherever possible, seasonal calendars along with activity (agricultural and water) calendars were prepared with both men and women of different communities to understand the key livelihood activities they were involved in. Further, agricultural and water activities in the study sites where men and women engaged individually and/or collectively were documented, along with the roles and responsibilities of men and women under each of these activities. This was further augmented with data from semi-structured household interviews.

Table 10.2 Study sites (*G: General Caste; OBC: Other backward caste; SC: scheduled caste); Source: Census, 2011

Village	District	Elevation Category (masl)	Total settlement area (Ha)	Total number of households	Population		Literacy		Major caste groups	Livelihood source
					Men	Women	Men (%)	Women (%)		
Hakeempur Tura	Haridwar	Plains (268)	282.88	419	1368	1278 (37%)	60.8	39.1	• Saini (OBC)- 26%; • Muslim (OBC)- 32%; • Harijan (SC)- 28%; • Kashyap (SC)-0.05%; • Valmiki (SC)- 0.02%	Primary: Agriculture Secondary: Daily wage labour (Census, 2011)
Kim Khola	Tehri Garhwal	Mid (1329)	117.5	95	170	243 (40%)	52.1	47.8	• Raiput (G)- 65%; • Harijan (SC) -35%	Primary: Agriculture Secondary: Daily wage labour;out-migration mainly in the Hospitality sector (Census, 2011 and primary data)
Huddu	Rudrapryag	High (1953)	89.19	68	180	181 (37%)	57.7	42.2	• Raiput (G – 75%); • Harijan (SC) - 24%	Primary: Agriculture Secondary: Daily wage labour, Religious tourism (Census, 2011)

Subsequently, across the study villages, 12 key informant interviews with village elders were conducted to collect perceptions of long-term climatic changes. Separate focus group discussions (18 in the plains and eight in each of the middle and upper hills) were organized for women and men from different ethnic groups (Table 10.2). The participants discussed the impacts of climate variability, asset ownership, social roles and relations within the community (Table 10.3). This information was tabulated for different activities to gauge differential access to assets in each site.

Study area

To understand the dynamics of gendered institutions, information was gathered across different locations in the Hindu Kush Himalayan region of India. A river basin approach to site location was followed, and for the purposes of the study our work was confined to the Upper Ganga Basin (UGB) falling within the Uttarakhand administrative boundaries. The elevation of the UGB ranges from about 7500m above mean sea level in the northern high mountains to about 100m in the lower areas of the basin (Bharati and Jayakody 2010). Given the large topographical variations resulting from these elevations, the climate in the state varies from the valleys to the slopes, from subtropical to temperate climate (Chauhan 2010). Moreover, Uttarakhand also witnesses stark inequality between its plains and hilly districts, with a higher proportion of women in its hilly districts, a trend substantiated by the growing out-migration of men from these areas, leaving behind women to manage everyday household work. For the present study, the UGB has been divided into three elevation regions – plains, middle and high-elevation zones (Figure 10.2). One village in each elevation was chosen to explore the different interplay of nested rules for agriculture and accessing water within the village. These villages represent regions that have witnessed climatic aberrations in the past. While agriculture is the predominant livelihood across these villages, there are differences in terms of gender relations between men and women of different communities.

The study sites fall within the district administrations of Haridwar, Tehri Garhwal and Rudraprayag districts. In the plains, Hakeempur Turra village in the Bhagwanpur tehsil of Haridwar district was selected. The village is situated at the foothills of the Shivalik mountains, where the river Ganga descends into the plains. It is located 31 km west of the district headquarters Haridwar. It is well connected to the industrial labour market of Haridwar. The village is approximately 1 kilometre from the Solani river, a tributary of the Ganga. The middle-elevation village, Kim Khola, is in Tehri Garhwal district. It is located roughly 7 km uphill from Devprayag and can only be accessed on foot. Kim Khola is located in the water-scarce Hindolakhal region, which has faced severe water scarcity for many years, adding to the vulnerability of the local people. The high-elevation case study site, Huddu, is a small village in the Ukhimath tehsil of Rudraprayag district. It is located 33 km north of Rudraprayag town, the district headquarters.

Table 10.3 Perceived climatic variability in plains elevation site

Climatic Parameter	Perceptions Hakeempur Turra (Plains)			
	Saini, Muslim Communities (OBC Category)		Kashyap, Harijan and Valmiki Communities (SC Category)	
	Men	Women	Men	Women
Ground Water	Declining depth	No observation	Declining depth	Declining depth
Erraticism in rainfall	Increased	No observation	Increased	Increased
Erraticism in events of hailstorms	Increased	No observation	Increased	Increased
Winter temperature	Increased	Increased	Increased	Increased
The intensity of summer squalls	Increased	No observation	Increased	Increased
Pest attack on crops	Increased	No observation	Increased	Increased

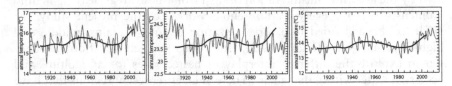

Figure 10.2 Yearly average temperature trend in Kim Khola, Hakeempur Turra, and Huddu respectively for 1901–2015.

Demographic profile

This section highlights the gender differentials in access and control over resources according to caste and class.

The caste system observed in the lower-elevation village, Hakeempur Turra, has been in existence for generations, based on the occupational status of particular communities. This allocation of duties depended mostly on the land holding and economic need of the household. The caste groups are present among both Hindu and Muslim groups. Large sections of the land are still mostly owned by the affluent Saini caste (and Gade within the Muslim community), while the other caste groups belonging to scheduled caste, Valmiki, Harijan and Kashyap communities (or Nai and Teli communities among Muslims) either work as agricultural labourers or as off-farm labourers in nearby factories or building works. According to the 2011 census, in Hakeempur Turra women comprise a mere 14 per cent of the total workforce. Field narratives suggested that in the census, women from the Harijan and Kashyap community households were identified as agricultural labourers as well as cultivators. Cultural taboos do not allow Muslim and Saini women to work in the fields, so that in the main worker category of the 2011 census, only 3 per cent of the total agricultural labourers are women. Muslim and Saini women were in charge of domestic household tasks, whereas the men were responsible for work outside, mainly agricultural activities.

In Kim Khola, women constitute about 66 per cent of the total workforce in the village. Around 38 per cent of the women were in the 'main worker' category and 70 per cent of them were 'marginal workers'. Of the marginal workers, 69 per cent of the women were cultivators, while 100 per cent of the women were agriculture labourers. In Huddu, of a total of 221 workers, 52 per cent were women., Women made up 76 percent of the marginal workers, while 80 per cent of the marginal worker women were identified as cultivators. Field observations revealed high rates of men's out-migration in both the high and mid-hill villages. There are 9.49 per cent, 31.09 per cent and 27.92 per cent of women-headed households respectively in Haridwar, Tehri Garhwal and Rudraprayag districts (SECC 2011). Other than widows, no woman in these three villages held land in her own name. From secondary data and our field observations, it is evident that women's role in agriculture increases with the elevation. This can be attributed to the high migration rates of men in the higher elevations compared to the plains of Haridwar (Bhandari and Reddy 2015).

Research findings and discussion

Prevalence of climatic variability in the upper Ganga Basin

The state of Uttarakhand lies in the country's highest seismic zone. It is vulnerable to geo-hazards and exposed to many types of hydro-meteorological extremes because of its geographical location. The region is made up of large stretches of land and rivers flowing down the mountains. Floods, cloudbursts, flash floods, glacial lake floods, hailstorms, water scarcity, drought, rockfalls, landslides, mud flows and forest fires are frequently reported incidents in one or other part of the state. The National Disaster Management Authority claims that the state is vulnerable to disasters. There have also been regular deluges and other disasters in the region in recent years (Kala, 2014). Over the years, floods originating from the Ganges river in Uttarakhand, for example, in 2010 (Bhatt and Rao, 2016) and 2013 (Sati and Gahalaut, 2013), have caused hardship for communities living both in the upstream and downstream regions. The 2013 flash floods in the Ganges affected many people and had serious consequences for local economies and livelihoods. While the nature of the events varied as we moved from site to site, it was found that these extreme events were not limited by altitude and occurred across all elevations with varied exposure to one or the other event.

The literature indicates a consensus that in recent years there has been a decrease in the number of rainy days and high variance in inter-annual rainfall, as well as changes in temperature trends leading to irregular weather patterns and increased climate uncertainty in the UGB. There is a clear trend of warming manifest through increases in the temperature at all altitudinal locations of the UGB. The high-elevation village, Huddu, was observed to be sensitive to spells of high rainfall, flash floods, landslides and mudflow, while the mid-elevation village, KimKhola, situated in the Devprayag block, experiences water scarcity. The low-elevation site, Hakeempur Tura, located near the river Song, gets flooded as the river expands when flows increase – especially its agricultural land. Figure 10.3 shows the annual average temperature from 1901 to 2015 collected by the Climate Research Unit, UK, in March 2017, for grids over the three villages of Hakeempur Tura, Kim Khola and Huddu. Although the data are available at a coarser resolution, the trend indicates an increase in temperature for all three villages, that is, the entire grid.

Aggregating grid-based rainfall data is difficult in the mountains due to elevation issues. The same grid may have multiple points of increase and decrease due to the valleys and the mountain peaks. This argument also holds true for temperatures. Hence, the rising temperature trends exhibited in the three locations are merely indicative. While there are no clear trends in precipitation, erraticism is reported to have increased from all the elevations, with reduced rainfall in the monsoon season. In certain cases, increases in intensity have also been reported, with increased incidences of cloudbursts, flash floods and floods. Landslides and mud flows have also increased over the years.

Figure 10.3 Water Resource Map for Hakeempur Turra Village.

In the following section this data is triangulated by combining it with the community's perception of climatic variability trends over the past few decades.

People's perception of climatic variability and its perceived impacts

The literature stresses that environmental and climate perceptions are constructed by people who identify threats and risks based on their way of understanding and perceiving them (Hopkins et al., 2001). Douglas and Wildavsky (1982) found that the understanding of climate change is a social process of specific threats and fears conditioned by social relations. Climate variability is identified and established through social and cultural factors (Kasperson et al., 1988). Wolfe (1988) states that climate change is experienced at least as much in relation to social factors as to quantified projections. Gender is considered a significant axis of differentiation within a society (Hopkins et al. 2001) and thus an important factor in climate vulnerability perceptions and experiences.

Owing to the differences in socio-economic backgrounds, exposure to risk varies considerably in each site across the elevations where the study was focused. Overall, communities pointed to changing trends in climatic parameters that have been witnessed in the recent past. Field interviews reveal that water stress, declining water availability, temperature changes and declining agricultural productivity are perceived in each site across the elevations on which the study was focused. Overall, communities pointed towards uncertainty in arrival of the monsoon, distribution of rainfall and increasing temperature.

Perceived climatic variability in the plains

The frequency and intensity of rainfall can be seen in the form of changing water availability, giving rise to issues of no water or excess water during the monsoon period. In a discussion with a Harijan community in Hakeempur Turra, a woman noted: "Earlier each spell of the monsoon showers would continue for three to four days. Nowadays the rainy days have decreased." Further, men from the Saini and Muslim communities often linked the falling levels of groundwater to "reduced overall rainfall" and correlated this with their increasing dependence on groundwater for irrigation and domestic purposes. Further, men farmers stated that temperatures have been rising during the winter but the intensity of summer squalls has reduced, and winter hailstorms have become erratic, adversely impacting standing wheat crops. A man from the Muslim community said: "The difference between what was perceived as summer and winter seasons has reduced." From discussions, it was clear that until the 1960s, farmers in the plains relied on soil moisture and surface water for farming on their fields and near the river bank. Now they rely on groundwater irrigation, using diesel or electric pumpsets. From the interviews and FGDs, men farmers from the Saini and Muslim communities in the plains showed greater awareness of the consequences of climate variability than the women. They mentioned the challenges they faced due to newer pests and diseases affecting the crops at different stages.

Perceived climatic variability in mid- and high elevations

In the mid-hills and high hills, the women farmers were more responsive than the men. They shared how uncertainty in rainfall is making them change sowing timings of their rainfed crops. A 70-year-old resident of Kim Khola, describing the climatic variability within his lifetime, said: "Fifty years ago, it used to snow for 8 to 10 days and the snow cover would last for at least 15 days. Now it does not snow at all. Rainfall has become erratic and the amount is lesser. Discharges have reduced to half the previous levels in two springs in our village. One spring that irrigated even small patches of land has completely dried up." In the middle-elevation region (Table 10.4), agriculture is largely rainfed. Changes in precipitation therefore impact agricultural livelihoods, with erratic rainfall damaging standing crops. In Kim Khola, villagers faced a water crisis due to the drying up of springs. The domestic water supply was interrupted for 10 days before intervention from external private and government agencies to augment the supply from newer sources. Women had to go 3,000 metres to collect water, which added to their workload.

Compared to the other elevations, agricultural lands in the low elevation were found to be more productive and certain crops like red kidney beans or vegetables produced good yields. The challenges of producing at economies of scale and target markets have, however, rendered agriculture largely

Table 10.4 Perceived climatic variability in mid-elevation site

| Climatic Parameter | Perception Kim Khola (Mid-elevation) | | | |
| | Rajput Community (General Category) | | Harijan (SC Category) | |
	Men (non-migrants)	Women	Men (non-migrants)	Women
Spring Discharge	Significantly Reduced	Significantly Reduced	Significantly Reduced	Significantly Reduced
Erraticism in rainfall	Increased	Increased	Increased	Increased
Snowfall events	Decreased, almost negligible	Decreased, almost negligible	Decreased, almost negligible	Decreased, almost negligible
Winter temperature	Increased	Increased	Increased	Increased
Pest attack on crops	Increased	Increased	Increased	Increased

Table 10.5 Perceived climatic variability in high-elevation site

	Perception Huddu (High-elevation)			
Climatic Parameter	Rajput Community (General Category)		Harijan (SC Category)	
	Men (non-migrants)	Women	Men (non-migrants)	Women
Spring Discharge	No effect	Decreased	No effect	Decreased
Hailstorms	Increased	Increased	Increased	Increased
Snowfall	Decreased	Decreased	Decreased	Decreased
Winter temperature	Increased	Increased	Increased	Increased

subsistence in the mid- and high elevations. A large part of the agricultural land is left fallow and is uncultivated. Falling productivity and damage to crops by wildlife have been highlighted as major reasons for abandoning agriculture, and have also been triggers for out-migration and shortage of labour in many cases. It is mostly men, including young boys, that have out-migrated. Due to a combination of the many factors highlighted above, women reported an increased dependence of households on market-based products, which had not been the case before.

In the high-elevation region (Table 10.5), there has been a decline in snowfall. Although at present this does not give rise to water stress, reduced recharging of aquifers over the next two to three decades may lead to water scarcity. The implications are being felt in terms of day-to-day water requirements. In Huddu, a retired 70-year-old man perceived "greater diurnal variation" in temperature: "It gets hot during the day. Often in the evenings the temperature falls considerably. We cannot be sure about the weather these days." Women farmers here perceived that average winter temperatures have risen compared to even two decades ago. In an FGD, some of them said: "Fans were not needed 20 years ago. The need was felt recently and table fans were introduced in most homes. Earlier we used light blankets in the summers too. Now blankets are used only at the onset of autumn." Moreover, because of the variability in temperature, farmers also believe that pest attacks have increased.

Gendered institutions: the action arena

Climate variability has intensified competition for water and agricultural resources. In this section, we explore the 'rules in use' and pattern of inter-actions in two action situations: (i) accessing water for domestic use; and (ii) agricultural activities.

Action situation 1: accessing water for domestic use

Water availability is changing in all three villages, with issues of both water scarcity and excess of water. It was observed that the interactions between men and women actors are governed by socio-cultural norms, especially along caste lines (Table 10.6). Interestingly, it was pointed out that caste and gender often intersect to influence actors' degrees of participation in accessing water, defining their social power. The source from which water is collected varies in most cases, reflecting differential vulnerabilities across social strata. The intra-household division of labour has been passed down the generations. In all study villages, it was mostly women from all socio-ethnic communities who performed the tasks of water collection from various sources and water storage for all domestic purposes.

The study village in the plains, Hakeempur Turra, displayed rigid conformism in water access, resulting in inequalities and social exclusion based on caste and gender. Each household in the village has traditionally depended on hand pumps at the depth of 55 ft. Villagers reported contaminated water from the hand pumps, suggesting a decline in water quality over the past two decades, making it unusable for domestic purposes. They attribute this decline in quality to multiple factors, including reduced rainy days leading to lower recharging of ground aquifers and rapid industrialization in the Bhagwanpur belt near the village.

In response the government of Uttarakhand installed about 30 government taps, randomly placed in the village, to cater to the domestic needs of all households. However, the participatory mapping exercise conducted in the village (Figure 10.4), revealed that these taps were concentrated around the Saini and Muslim community households (regarded as affluent castes) and sparsely scattered around the Harijan, Kashyap and Valmiki communities (the poorer/lower castes) (Table 10.6). This implied that women from Harijan, Kashyap and Valmiki households had to travel longer distances and spend more time than women from the Muslim and Saini communities fetching potable water on a daily basis. On further enquiry, it was noted

Table 10.6 Distribution of potable water source in Hakeempur Turra

Community		No. of households	No. of government water stand post	No. of private taps (private submersible pumps)
Muslims (OBC)	Gade	120	10	6
	Teli	11	3	0
	Nai	7	2	0
Saini (OBC)		150	12	22
Kashyap (SC)		22	4	2
Harijan (SC)		120	1	2
Valmiki (SC)		9	1	0

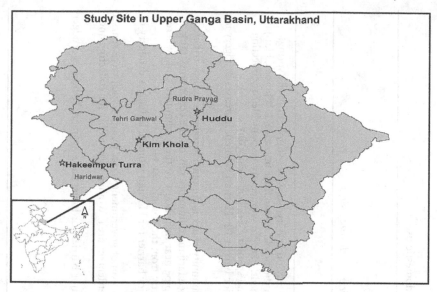

Figure 10.4 Map depicting the study sites in the high, mid- and plain regions of Uttarakhand.

that power dynamics had played a major role in deciding the placement of the government taps, hinting at inadequate opportunities for lower-caste communities as compared to the affluent castes (Table 10.7).

Apart from the government water taps, a few individual households had personal submersible water pumps at a depth of 250 ft. These personal submersible water pumps, which required an investment of Rs. 30,000, were mostly owned by Saini and Muslim households. In the case of unavailability/lower discharge of water from government taps in the village, women from Harijan, Kashyap and Valmiki communities are often not allowed to share the water sources located in Muslim or Saini neighbourhoods. However, sometimes children from these communities buy water from Muslim or Saini private submersible pump at locally determined fixed rates.

Similar trends were observed in the mid-elevation village Kim Khola, where caste often becomes a determinant of gender roles, adding to women's workloads. The village is affected by water scarcity and reduced numbers of rainy days resulting in reduced soil moisture, infiltration, recharge and discharge in springs. During interviews, it was noted that there are four perennial springs in the village used for domestic and livestock-rearing purposes. These springs used to be the primary source of water for domestic purpose. On further discussion, however, villagers mentioned that since the 1980s, the discharge from all the springs has drastically declined while one of them has dried up completely. The springs have now been replaced by two overhead tanks with water supplied from the Bhagwan Pumping Scheme in the Devprayag cluster. The Harijan caste relies on the community water tank, which is separate from the tank serving the Rajputs. For

Table 10.7 Rules and implications for actors in action situation 1: accessing water for domestic use

Action Situation: *Accessing Water for Domestic use*
Activity: *Fetching Water for Domestic use*
Actors: *Men and Women from different communities in each study site*
General Rule-in-Use: *Women performed the task of water collection from all communities in all study sites*

Study Village	Ethnic Groups	Rules and Implications
Huddu Village, High-elevation	Rajput (G) and Harijan (SC)	**Rules-in-use in the institutional setting:** Rajput community rely on gravity led water supply from the spring. Women responsible for collecting and storing water from the pipeline near their household. Whereas, Harijan community rely on community spring uphill their settlement. Harijan women are responsible for collecting water directly from the spring by travelling 7km uphill and spending 5-6 hours each day. **Implication for the Actor:** Differential water access is a major barrier for Harijan women. Increases drudgery in case of scarcity
Kim Khola Village, Mid-elevation	Rajput (G) and Harijan (SC)	**Rules-in-use in the institutional setting:** For the Rajput community water is supplied through a well-distributed network of gravity based piped community taps while the Harijan community rely on the community tank. Rajput women collect water either from the community pipe near their house or the community tank, whereas, Harijan women are responsible for collecting water directly from the community tank. **Implication for the Actor:** Differential water access is a major barrier for Harijan women. Enhances drudgery in case of scarcity
Hakeempur Turra Village, Plain elevation	Saini, Muslims (OBC), Harijan, Kashyap and Valmiki (SC)	**Rules-in-use in the institutional setting:** Differential distribution of government water pumps which were sparsely scattered in Harijan, Kashyap and Valmiki community settlement area. Women responsible for collecting and storing water from all households. **Implication for the Actor:** Differential Water Access is a major barrier for lower caste women. Enhances drudgery in case of scarcity.

the Rajputs, the tank supplies water through a well-distributed network of gravity-based piped community taps, as the Rajput settlement is downhill from the tank. The Harijan settlement, being uphill from the water tank, lacks a gravity-enabled network of community taps. Although the distance from the Harijan settlement to their water tank is less than 2 km, Harijan women need to make this trip to access water, while Rajput women have access to water source in their home. If the water supply is affected, these Harijan women need to journey nearly 4 km to collect water, spending nearly 4–5 hours of every day on water collection. Harijan women reported that the task of collecting enough water from the community tank to meet the requirements of the whole family is exhausting, given the low yields during the summer months.

Similar caste differentials in water access were noted in the high-elevation village Huddu due to location of the settlements vis-à-vis the springs. The Rajputs live uphill of the Harijans. Traditionally, water is supplied through pipelines from separate springs to both communities. The Rajput spring is located above their settlement and gravity takes the water to each household through the pipeline. The Harijan spring is at a lower elevation than the settlement, so the pipeline does not function. The Harijan women have to spend 5–6 hours every day on a 7-km uphill walk to collect water directly from the spring. On further questioning, Harijan women stated they need external support to install diesel pumps in their settlement area to ease this drudgery..

Action situation 2: agriculture activities

Agriculture was observed to be the primary source of livelihood across all sites. It was informally noted that division of labour within agricultural activities varied between different communities. The dynamics of each community were complex and discrete in terms of the nature of the work they performed. The different roles within the household were shaped by several factors that in some cases restricted the women's role, and hence had an impact on their vulnerabilities.

In the low-elevation village, Hakeempur Turra, the division of labour largely assigned women to the reproductive sphere of the home, while men were engaged in the public sphere of production and employment among the OBC communities. This includes all agricultural work and dealing with the market (Table 10.8). They plan the crop cycle and take all the major farm decisions. In Muslim households, women are tied to the house and perform reproductive chores. Though some younger women have started moving out of the village for higher education and technical jobs nearby, FGDs revealed that they were mostly unaware of the location of their fields. Within the SC communities, however, there is greater diversification of livelihoods. Lower dependence on agriculture reduces their vulnerability to climate shocks. Most SC landholdings are less than 10 bighas (1 bigha = 0.4 acres), and women and men are equally involved in farming activities

Table 10.8 Rules and implications for actors in the high hills, mid hills and plains in action situation 2: agriculture-related activities (✓ *rule in use, *non-migrants, # applicable only for plains, Highlighted cells shows activities taken by men and women together*)

Action Situation: Agricultural Activates
Activities: Preparation of land (ploughing); Sowing, Plantation; Weeding; Leasing land; Harvesting; Irrigation; Application of water on fields; Saving Seeds; Buying seeds, fertilizers and pesticides#; Buying seeds; Adoption of new technology; Access to telecommunication for farming
Actors involved: Men and Women farmers from Rajput and Harijan communities in Huddu and Kim Khola village and Men and Women farmers from Saini, Muslim, Harijan and Kashyap communities in Hakeempur Turra
Perceptions on climatic variability: Decreased land productivity, changing cropping pattern, reduced rainy days
General Rule-in-Use for high and mid elevation: Farming is a woman centric activity; Men are only engaged in land preparation activities.
General Rule-in-use for plains: Division of labour within the wider economy assigned women to the reproductive sphere of the home and everyday farmwork while men were allocated to the public sphere of the production and employment. but in marginalized communities, both men and women take up equal roles for income generation.

Action Arena: Agricultural Related Activity	High elevation				Mid-elevation				Low elevation			
	General (Rajput)		SC (Harijan)		General (Rajput)		SC (Harijan)		OBC (Saini and Muslim)		SC (Harijan and Kashyap)	
	Men*	Women	Men	Women	Men	Women	Men	Women	Men	Women	Men	Women
Preparation of land (ploughing)	✓		✓		✓		✓		✓		✓	
Sowing		✓		✓		✓		✓	✓	✓	✓	✓
Plantation		✓		✓		✓		✓	✓	✓	✓	✓
Weeding		✓		✓		✓		✓	✓	✓	✓	✓
Leasing land	✓		✓		✓		✓		✓		✓	
Harvesting		✓		✓		✓		✓	✓	✓	✓	✓
Irrigation, Application of water on fields		✓		✓		✓		✓	✓	✓	✓	✓
Saving Seeds		✓		✓		✓		✓	✓		✓	
Buying seeds, fertilizers and pesticides#	✓		✓		✓		✓		✓		✓	
Adoption of new technology, Access to telecommunication for farming	✓				✓				✓		✓	

on their own land or in the fields of the OBC families. Consequently, SC women acquire productive skills in addition to reproductive ones, increasing their say in decision making.

Agriculture in the mid- and high hills is largely subsistence in nature, unlike in the plains. The majority of people depend on agriculture as their main livelihood, along with construction and labour work. The land is largely fragmented and scattered with very small landholdings, making it challenging for the communities to produce enough for the markets, still less to produce surpluses. Men here have migrated, whereas women continue with agriculture, which requires a major labour contribution from women throughout the year. Men only plough the fields using bullocks or help with harvesting if they are home. When the men are away, women hire labour to plough the fields. Women's roles in households with migrants and those without are almost the same. It was observed that men out-migrants interviewed during their return to the village for holidays had little awareness of climate variability and its impacts on natural resources in the area.

Fodder collection is also one of the primary activities for women a considerable amount of time is spent daily carrying fodder long distances on their heads, backs and shoulders in addition to fetching water and firewood. As well as increasing their workload, this increases ther physical burdens. Livestock was once a steady source of income. However, due to out-migration and reduced availability of fodder, families have reduced the number of domesticated animals they keep. The reduction in agricultural yields, male out-migration, stress and workload seem to have increased among women over the years. Lower-caste communities were found to have less land holding and therefore, apart from working as agricultural labourers, these families mostly engage in off-farm livelihoods such as building and other forms of casual labour.

The cropping system is traditional, and unlike in the plains, decisions regarding changes in crop patterns are taken jointly by women and men, often even solely by women if men have out-migrated. Overall, land tenure is associated with men in all elevations, so decisions related to the lease and exchange of land are made by men, though in consultation with their women. Land or any asset is in the name of a woman only if she is widowed, although in this situation the final decision would be taken by the eldest son of the family. Women in this region are heavily involved in agriculture, as in other mountain states – except for ploughing the fields, culturally associated with men. However, with the growing out-migration (predominantly of men) witnessed in the villages of Uttarakhand (Ghosh et al., 2007) this activity is also now being performed by women.

To sum up the action situation, distinct divisions were seen in different communities, resulting in a lack of skill development for women in a few communities. In the lower-caste households, due to smaller land holdings and major economic need, women also assisted male family members in the field and worked as labourers. This resulted in a higher workload but the acquisition of more better skills in productive activities along with their household duties. Similar results were found in a study conducted by Djoudi and Brockhaus (2011). Due to the higher involvement of men in agricultural activities in the plains, they had greater powers of decision in the household.

Exploring gender relation patterns in the three elevations

As discussed in the previous sections, both action situations (accessing water and agricultural activities) involved interactions between women and men and were often influenced by availability of and access to assets, which in turn were governed largely by the actor's perception and the intrinsic rules in use for each study site. It was also noted that climate variability impacts each action situation. In this section, we evaluate the key patterns emerging from interactions within the gendered institutions across the elevations.

Gendered institutions: dynamic, complex and diverse?

Gender shapes social relations and situations in profound ways. Through the institutions of norms, rules and values, gender becomes an underlying structure embedded in everyday interaction and expressed in "perceptual, interactional and micro-political activities" (West and Zimmerman 1987). Narratives from the field suggest that women and men are a heterogeneous group, with overlapping ethnic, class and caste identities that result in multiple forms of marginalization and exclusion. An exclusive focus on women glosses over differences between women and their interests. Caste and gender often intersect to influence individuals' degree of influence in the community, or in short, their social power. Addressing gender requires an understanding that women (as well as men) are not a homogeneous category and the continued assumptions made around a single homogeneous class of 'mountain' women would reflect an incomplete picture. This highlights the dynamic, diverse and complex nature of gendered institutions.

Gendered rules, norms and rights: rigid, flexible or fluid?

The intra-household division of labour plays a key role in defining vulnerabilities to climate change. Often, within the literature, the division of labour generates gendered processes that are classified into separate 'reproductive (private)' and 'productive (public)' spheres, characterized by specific priorities and privileges. This results in women being assigned to the reproductive sphere of the home and everyday farm work while men are involved in the public sphere of production and employment, with differential access/rights to livelihood resources. Field narratives depict that in the high and mid-hills, due to the high out-migration of men over the years and the higher involvement of women in agricultural activities, women make the immediate decisions regarding the crops to be grown. But in the plains, due to higher caste conformity, women from OBC communities were less empowered to make decisions and had little knowledge about agricultural activities, since they were only associated with household responsibilities. This highlights that social and human-environmental interactions often challenge gendered norms, rules and values. While some boundaries and relations seem to be rigid, others are more flexible or fluid. Further, institutional networks prefer men and assume them to be heads of households. Interestingly, the position of village head (*Pradhan*) in the plains and high hills is reserved for SC women. This should imply that the *Pradhan* represents the voice of her community's women and introduces measures that might reduce their drudgery. Interviews with them revealed, however, that major decisions regarding natural resource management and livelihoods are made by either the elected *Pradhan*'s husband or father-in-law. In accordance with social norms, a man in her family is recognized as *Pradhan* by society, even when the actual position has been won by the woman, limiting her ability to voice her opinions. There was also a difference in the participation of women in meetings held at the village institution (Panchayat) level. In the mid- and high hills, when asked about their dependency on groundwater/springs, women expressed their wish to revive the community water harvesting pond but also expressed their inability to do so, given the power dynamics. Further, no women's self-help groups (SHGs) were found in the high and mid-hills study sites. Only one SHG was identified in the plains, in which only of women of the Kashyap caste (SC community) participated. This highlights the power dynamics between different caste groups: more static in the plains and more dynamic in the mountains.

Social relations, position and mobility: inclusive, partially inclusive or exclusive for all?

People are positioned differently in terms of their access to and control over assets. While some have direct control, others access these through mediators, often requiring extra effort. This shapes women's and men's capacities, incentives and preferences for how to access, use and control the assets defining their social position. Narratives presented in the previous sections show evidence of vulnerability and its impact on people lower in the socio-economic hierarchy, irrespective of their elevation. In the mid- and high hills, differential water access was seen that indirectly resulted in increased drudgery for women from SC households, while in the plains, we found no uniform distribution of government taps. This highlights the need to focus on changing gender relations around water to yield a better understanding of the impacts of drinking water supply interventions on women's quality of life, rather than estimating impact simply in terms of conventional indicators of women's water burdens.

Further, it was observed that women across all communities acquire agricultural land through men (i.e., mainly husbands) who inherit, own and control land. It was also observed in the plains that for educational purposes, the men of the family may be sent outside the village to attend university or diploma courses, which increase their chances of getting jobs outside farming. Women attend educational institutions within a 3–4-km radius of the village and are often married off after their education. However, they make take up jobs like teaching within the village. This trend was also observed in the high and mid-hills sites, where men belonging to socially dominant castes, having better access to education, financial resources and social networks, were more likely to migrate. Further, differential labour force participation and wages was seen at all elevations, with women employed in private construction earning lower pay than men for the same job role (Rs. 250/day for women and Rs. 300/day for men). This again highlights that social relations, position and mobility might not be inclusive for all, and differ in different social, political and economic settings, dynamically shaping "gender as a critical variable in shaping resource access and control" (Rocheleau et al., 1996).. However, how social position and mobility will play out in terms of climate change is yet to be fully seen and was not explored through this research.

Willingness to learn, evolve and adapt to climatic variability: intrinsic or socially contingent?

Operational rules that define the roles and responsibilities of men and women have been intrinsic to gendered institutions. These rules have been evolving very gradually in all the elevations, with men and women starting to diversify their livelihoods to cope with the changing climate. From field observations it was noted that operational rules are interwoven with the cultural mesh of the system that defines the level of conformism to caste norms and customary principles. Therefore, evolution at the level of collective choice rules and constitutional choice rules will always hinge on the ability of men and women from communities to resist and move away from, rather than conforming to, the operational rules. Accordingly, in the high and mid-hills, with an increase in men's out-migration, women have a greater say in both productive and reproductive activities and find themselves more aware within their households. But differential access based on caste still exists, which shows that the strains of social conformism are still strong. In the plains, the degree of conformism is greater, meaning that any change in the operational rules will take place at a much slower pace. Therefore, in all sites, the actual empowerment of the lower-caste (SC) communities is very slow, hinting that willingness and social conformity should be seen in unison in order to represent a more realistic picture.

Conclusions

This chapter has focused on the impacts of climate variability on men and women engaged in accessing water for domestic purposes and in agricultural activities, using the IAD framework. Observations demonstrated that men and women across diverse communities and elevations interpret, experience and respond to climatic and non-environmental changes concurrently, yet differently.

Differentiated impacts of climate variability within gendered institutions were manifested that are due to unequal distribution of roles, responsibilities and resources for men and women from different communities in all study sites. These factors had a direct impact on their assets, their access to livelihood attributes and the options available to them for pursuing optimal livelihood strategies. Moreover, the gendered division of labour between men and women reflected a strong dichotomy between 'reproductive' and 'productive' work in all the study sites. It was noted that the actors whose livelihoods are linked with natural resources are more confident to report on the changing biophysical parameters and climate-related awareness. It can be said that differential water access for domestic use made women, especially those from SC communities across all elevations, more exposed to the impacts of climate variability. In the high and the mid-hills, due to men's out-migration in search of productive economic activities, women from all communities found themselves performing both agricultural activities and household work. In the plains, women from OBC communities were tied to the home and had restricted mobility, while SC women were better equipped with livelihood diversification skills, as they were involved in both productive and reproductive activities. Socio-cultural conformism was seen as the prime cause limiting the voice of OBC women in decision making, education and participation in formal and informal institutions.

The institutional perspective to understanding gender dynamics facilitated our ability to bring out the gendered differences in social institutions on the basis of the intrinsic socio-cultural norms across elevations, as all the study sites are within the same state-level administrative boundary. Moreover, our analysis has emphasized the need to take account of both the characteristics these institutions exhibit and the necessity of situation-specific climate change interventions. To gauge the resilience of these gendered institutions to climate change, inherent properties of the system like equity and adaptability can provide an appropriate scale for deepening our understanding of the context, thereby improving the design and implementation of climate change-related interventions in the future.

Note

1 This work was carried out by the Himalayan Adaptation, Water and Resilience (HI–AWARE) consortium, under the Collaborative Adaptation Research Initiative in Africa and Asia (CARIAA), with financial support from the Department for International Development, United Kingdom (DfID), and the International Development Research Centre, Ottawa, Canada.

References

Adger, W.N. 1999. Social vulnerability to climate change and extremes in coastal Vietnam. *World Development in Full*, 27(2): 249–269.

Aguilar, L. 2009. Women and climate change: Vulnerabilities and adaptive capacities. *World Watch Institute*, 59(80): 10–29.

Ahmed, S., and Fajber, E. 2009. Engendering adaptation to climate variability in Gujarat, India. *Gender and Development*, 17(1): 33–50.

Arora-Jonsson, S., 2011.Virtue and vulnerability: discourses on women, gender and climate change. *Global Environmental Change*, 21(2): 744–751.

Bhandari, G., and Reddy, B.V.C. 2015. Impact of out-migration on agriculture and women work load: An economic analysis of hilly regions of Uttarakhand India. *Indian Journal of Agricultural Economics*, 70(3): 395–404.

Bharati, L., and Jayakody, P. 2010. *Hydrology of the Upper Ganga River*. Project report submitted to WWF, India.

Bhatt, C.M., and Rao, G.S. 2016. Ganga floods of 2010 in Uttar Pradesh, north India: A perspective analysis using satellite remote sensing data. *Geomatics, Natural Hazards and Risk*. Taylor & Francis: 747–763. doi:10.1080/19475705. 2014.949877.

Björnberg, K.E., and Hansson, S.O. 2013. Gendering local climate adaptation, local environment. *International Journal of Justice and Sustainability*, 18(2): 217–232.

Brody, A., Demetriades, J., and Esplen, E. 2008. *Gender and Climate Change: Mapping the Linkages – A Scoping Study on Knowledge and Gaps*. BRIDGE, Institute of Development Studies, Brighton

Byg, A., and Salick, J.2009. Local perspectives on a global phenomenon—Climate change in Eastern Tibetan Villages. *Global Environmental Change*, 1(19): 156–166.

Chaudhary, P., and Bawa, K.S. 2011. Local perceptions of climate change validated by scientific evidence in the Himalayas. *Biology Letters*, published online, 27 April 2011; doi:rsbl.2011.0269.

Chauhan, M. 2010. A perspective on watershed development in the Central Himalayan State of Uttarakhand, India. *International Journal of Ecology and Environmental Sciences*, 36(4): 253–269.

Di Gregorio, M., Fatorelli, L., Paavola, J., Locatelli, B., Pramova, E., Nurrochmat, D.R., May, P.H., Brockhaus, M., Sari, I.M., and Kusumadewi, S.D., 2019. Multi-level governance and power in climate change policy networks. *Global Environmental Change*, 54(Jan): 64–77.

Djoudi, H., and Brockhaus, M., 2011. Is adaptation to climate change gender neutral? Lesson from communities dependent on livestock and forest in Northern Mali. *International Forestry Review*, 13(2): 123–135.

Douglas, M., and Wildavsky, A. 1982. *Risk and Culture*. Berkeley; London: University of California Press.

Ghosh, N., Kar, S., and Sharma, S. (2007). *Inequalities of Income Opportunity in a Hilly State: A Study of Uttarakhand*. IEG Working Paper Series No. E/287/2007.

Hopkins, W.D., Fernandez-Carriba, S., Wesley, M.J., Hostetter, A., Pilcher, D., and Poss, S. 2001. The use of bouts and frequencies in the evaluation of hand preferences for a coordinated bimanual task in Chimpanzees (Pan troglodytes): An empirical study comparing two different indices of laterality. *Journal of Comparative Psychology*, 115: 294–299.

Immerzeel, W.W., van Beek, L.P.H., and Bierkens, M.F.P. (2010). Climate change will affect the Asian water towers. *Science*, 328(5984): 1382–1385. doi:10.1126/ science.1183188.

IPCC. 2007. Climate change 2007: The physical science basis. Contribution of Working Group I to Solomon, S., Qin, D., Manning, M., Chen, Z., Marquis, M., et al., (eds), *The Fourth Assessment Report of the Intergovernmental Panel on Climate Change*. Cambridge, UK and New York, USA: Cambridge University Press.

IPCC. 2014. Climate change 2014: Impacts, adaptation, and vulnerability. Part A: Global and sectoral aspects. Contribution of Working Group II to Field, C.B., V.R. Barros, D.J. Dokken, K.J. Mach, M.D. Mastrandrea, T.E. Bilir, M. Chatterjee, K.L. Ebi, Y.O. Estrada, R.C. Genova, B. Girma, E.S. Kissel, A.N. Levy, S. MacCracken, P.R. Mastrandrea, and L.L. White (eds), *The Fifth Assessment Report of the Intergovernmental Panel on Climate Change*. Cambridge, UK and New York: Cambridge University Press.

Kala, C.P. 2014. Deluge, disaster and development in Uttarakhand Himalayan Region of India: Challenges and lessons for disaster management. *International Journal of Disaster Risk Reduction*: 143–152. doi:10.1016/j.ijdrr.2014.03.002

Kasperson, R., Renn, O., Slovic, P., Brown, H., and Emel, J. 1988. Social amplification of risk: A conceptual framework. *Risk Analysis*, 8(2): 177–187.

Lambrou, Y., and Laub, R. 2004. *Gender Perspectives on the Conventions on Biodiversity, Climate Change and Desertification*. United Nations, Geneva. Gender and Population Division.

Lowndes, V., and Roberts, M. 2013. *Why Institutions Matter*. Basingstoke: Palgrave.

Neumayer, E., and Plümper, T. 2007. The gendered nature of natural disasters: The impact of catastrophic events on the gender gap in life expectancy, 1981–2002. *Annals of the Association of American Geographers*, 97(3): 551–566.

North, D. 1990. *Institutions, Institutional Change and Economic Performance*. Cambridge: Cambridge University Press.

Ogra, M.V., and Badola, R. 2015. Gender and climate change in the Indian Himalayas: Global threats, local vulnerabilities and livelihood diversification at the Nanda Devi biosphere reserve, *Earth System Dynamics*, 6(2): 505–523.

Orlove, B., Roncoli, C., Kabugo, M., and MAjugu, A. 2010. Indigenous climate knowledge in Southern Uganada: The multiple components of a dynamic regional system. *Climate Change*, 100(2): 243–265.

Ostrom, E. 1990. *Governing the Commons: The Evolution of Institutions for Collective Action*. New York: Cambridge University Press.

Ostrom, E., Gardner, R., and Walker, J.M. 1994. *Rules, Games, and Common-Pool Resources*. Ann Arbor, MI: University of Michigan Press.

Pellicciotti, F., Bauder, A., and Parola, M. (2010). Effect of glaciers on streamflow trends in the Swiss Alps. *Water Resources*, 46(W10522). doi:10.1029/2009WR009039.

Rocheleau, D., ThomasSlayter, B., Wangari, E. 1996. *Feminist Political Ecology: Global Issues and Local Experiences*. London and New York: Routledge, xviii; 327.

Sati, S.P., and Gahalaut, V.K. 2013. The fury of the floods in the north-west Himalayan region: The Kedarnath tragedy. *Geometrics Natural Hazards Risk*: 193–201. doi:10.1080/19475705.2013.827135.

Scott, J. 1986. Gender: A Useful Category of Historical Analysis. *American Historical Review*, 91: 1053–1075.

Singh, S.P., Bassignana-Khadka, I., Karky, B.S., and Sharma, E. (2011). *Climate Change in the Hindu Kush-Himalayas: The State of Current Knowledge*. Kathmandu: ICIMOD.

Vedwan, N., and Rhoades, R.E. (2001). Climate change in the western Himaalayas of India: A study of local perception and response. *Climate Research*, 19: 109–117. doi:10.3354/cr019109

West, C., and Zimmerman, D. 1987. Doing Gender. *Gender and Society*, 1(2), 125–151. Retrieved 16 November 2020, from http://www.jstor.org/stable/189945

Wolfe, A.K. 1988. Environmental risk and anthropology. *Practical Anthropology*, 10 (4): 4.

11 Shaping gendered responses to climate change in South Asia

Asha Hans, Anjal Prakash, Nitya Rao, and Amrita Patel

Introduction

The purpose of *Engendering Climate Change: Learning from South Asia,* as explained in the introduction to the volume, was to explore and contribute to the climate discourse. Climate change is a historically shifting conversation and the knowledge which emerged in this volume is specific to the gendered understanding of the climate change hotspot study sites of South Asia. In this area, the vulnerabilities created by climate change have been challenged by women and men, sometimes successfully, and at other times, as observed in these writings, with an increase in vulnerabilities. Globally and in South Asia, policies have reflected the intention of meeting climate targets, yet the conversations on climate change have not been gendered. In recent years the situation has begun to improve as a debate with a gendered underpinning is being built globally and in South Asia to understand the vulnerabilities faced by women (Rao et al. 2019; Fernandes 2018; Ngigi et al. 2017; Goh 2012; Ahmed and Fajber 2009).

An evolving gendered framework for climate change research

Three key insights emerge from our discourse. The first is the importance of attention to inter- and intra-household relations in research. The analysis in the various chapters is based on lived experiences of women who are vulnerable to climate change. It reveals how their roles change within and outside the household in terms of relationships and structures. Within this framework we see how the household responds to risk and adapts to the stresses created by climate change. Second, this research establishes an improved understanding of the gendered impacts of climate change and risk management through adaptive strategies undertaken by local communities. Third, we explore how vulnerabilities are gendered, and agency is constructed, despite the existence of social and political barriers across the states of South Asia.

Based on these arguments, the analytical framework assumes that alternate future pathways require gendered policy responses. These need to include the costs of adaptation efforts and strategies through increased financial allocations and research on issues critical to gender equality and climate justice in South Asia.

Power across geographical locations: inter- and intra-household relations

The writings in this volume provide a gendered conceptual lens on changes taking place within the household in the South Asian region. What emerges from the research enquiry is grounded on the assertion that "climate policy is not a grand global narrative, but rather a series of small-scale decisions made at various scales that affect individuals in disparate ways" (Bee et al. 2015:6). This insight gives an understanding of how the gendered social positions and power dynamics within the household intersect with geographical and climate variabilities. The research highlights the contexts and correlations in intra- and inter-household structures and gendered strategies adopted to manage risk. Intra-household choices have depended not only on the structure and composition but also the socio-cultural positioning of women and men within the household.

A basic highlight has been the recognition of gender diversity within the household, that it varies by age, ethnicity, economic class and geographical location. Across the various chapters, evidence is presented on changing gender roles and the sharing of labour and assets. To obtain this evidence authors have engaged with households through the use of different methodologies, but they all take a critical lens that goes beyond the binary understandings of gender (Singh, participatory and Hazra, Vincent, quantitative survey). These investigations, which have been related to how gender is inscribed in the household, explore its multi-faceted dimensions and the diverse ways in which intersectionality is constructed and enacted in practice.

In this context we can use Seager's argument that globally, climate change has to be viewed from a perspective of "privilege, power and geography" (Seager and Olson 1986). From a feminist lens too, privilege and power are two important constituents of patriarchy which exist across South Asia (White 2017; Chowbey 2017; Sultana 2012; Isran and Isran 2012; Kabeer 1997; Kandiyoti 1988). This is a significant dimension, as we have discovered that while women exercise individual agency in responding to climate change, their reactions are located within socio-cultural contexts of patriarchy.

Despite the historical resistance to women's exclusion and abuse, patriarchy has not found a place in climate change research even when the issues are feminized. This is because of, as Enloe argues, "a lack of feminist curiosity" (Enloe 2017: 164). It is by increasing our curiosity about the household in climate hotspots, increasing our understanding and acting on it, that we can dismantle it. Gilligan and Snider argue that this is imperative as it is "powerful in shaping how we can see the world, that it can literally keep us from seeing what is right" (2018: 38). This script is familiar as it exposes the history of gendered relations in their narrow template that does not demonstrate interconnectedness outside the binaries.

The shifting dynamics of this gendered binary can be observed in the changed roles: it was found that men who do not migrate do share household burdens, for example, fetching water, perceived as women's work, and that women assume new gendered roles in agriculture (Solomon and Rao, this volume). Limiting ourselves to a binary enquiry would, as what Gilligan and Snider have suggested, lose sight of the multi-dimensional identities of women as mothers, workers, displaced or migrants, thus overlooking their differential roles. Findings across the chapters reveal that women's vulnerability and agency is shaped in inter-household and intra-household contexts, and related to their ability to access material assets and markets, work and develop adaptive strategies seeking to diversify their livelihoods. Some of the chapters demonstrate that households headed by women, including widows, have made inroads into the male locations of power.

Caste plays an important role in households, and Scheduled Caste women in India tend to be highly vulnerable in terms of access to water, for instance. Yet they do emerge with enhanced power when equipped with livelihood diversification skills (Rijhwani in this volume). Widows, as expected, are highly vulnerable but do overcome some adversities with the support of NGOs and the government, leaving non-widows relatively poorer and with higher workloads and responsibilities (Hazra et al. in this volume). This differentiation is again highlighted when older women with higher involvement in household decision-making are still not found to be the final decision makers, as men still dominate crucial adaptation decisions, especially those related to agriculture. This is not only a result of a patriarchal culture of control over assets and resources by men but of patriarchal government policies that target only men for agriculture training and extension programmes (Qaisrani and Batool in this volume).

Gendered migration research reveals that the male–female binary in the household intersects with those of labour markets (Singh 2019; Chant 1998; Lawson 1998). Significantly in this context, patriarchy emerges in restricting women's rights to migrate; for instance in Pakistan, women's mobility was observed to be restricted to avoid negative gossip about them, and when husbands are away young newly married women face further restrictions (Qaisrani and Batool in this volume). In both migrant and non-migrant locations, women's work is increasing, as work at home is combined with newer livelihoods ranging from animal husbandry to mill work. There are other critical issues such as the loss of women's agency when their personal assets, such as gold, are taken by husbands to pay for digging borewells (Solomon and Rao in this volume).

A contextual analysis, as we attempt in this book, contributes to a broader understanding of vulnerability derived from social discrimination and political inaction, while also pointing to the small but significant ways in which women are negotiating unequal power relations. Gendered changes due to the changing climate go beyond the binary understanding of gender and are inclusive of caste, class and ethnic diversity. Women from wealthier families, for instance, do not work in agriculture, though they might still be responsible for domestic water collection. The chapters draw attention to the often invisible role of intra-household decisions in enhancing vulnerabilities, and to attempts by women to break patriarchal barriers by evolving new gendered roles or engaging with collective action. This helps us to understand the relationships between the drivers of patriarchy and climate change, whether it is in the selling of gold to buy a water source or restrictions on women's mobility.

Future research needs to further amplify this situation to enable a better understanding of patriarchal control over decision-making, while at the same time uncovering changes in men's roles where these are taking place. Women within households are contributing to the creation of new approaches and techniques to meet the challenges posed by climate change, and this needs to be recognized both in research and policy.

Communities and gendered impact of climate change: risk management through adaptive strategies

Tracing common gendered trends across countries and gendered strategies adopted across borders has provided critical knowledge of gendered transformation at the community level. In the four countries in this volume, women's contribution to the development of climate resilience can be seen in their robust response to the adverse impacts of climate change.

Specific strategies for each type of hazard are required in South Asia, as there are storm surges, cyclones, flooding, drought, sea-level rise, melting snow and landslides, as witnessed in the chapters in this volume. These changes traverse semi-arid zones, mountainous and delta regions. Heat has emerged as in important issue with which women have to cope, reducing their work output and creating health issues (Abbasi et al. in this volume). The learnings from various authors exploring gendered responses to climate change are reflected in their attempts to shift patterns of behaviour. The scholarship is well defined, for instance as a consequence of climate change women's involvement in agriculture changes, often increasing, with women taking on additional roles as their men migrate in search of other work. Yet given their limited ownership of land, productivity and yields remain relatively low, particularly in contexts of climate change and food insecurity (Goodrich et al. 2019; Rao et al. 2020). With declining agricultural yields, women cope by shifting to cash crops, crop diversification and the adoption of climate-resilient crops, all of which boost their purchasing power and food security. However, it is usually men who remain the decision makers and carry out technology-related work. By relegating cash crops as an enterprise under men's purview, society restricts women's role in decision-making, while increasing demands on their time (Solomon, Rao in this volume).

One of the major contributions of the volume is the study of climate change related to water availability and differentiated water use practices in India that are impacting men and women within the household as well as the community. Water in South Asia has been a communal property (Mangi et al. 2019; Akter et al. 2017; Price et al. 2014). Its usage shows a shift from communal to individual control where purchasing power defines the gendered positioning, with women having less purchasing power. These impacts are differentiated across levels with shifts in work burdens within households, the erosion of traditional institutions at community level, and over-extraction and growing reliance on private water at a watershed level. For example, people with borewells can access water for irrigation and those with purchasing power can buy drinking water from tankers. Water shortages therefore have equity implications as they disadvantage the most marginalized and exacerbate existing vulnerabilities (Singh in this volume). Within households, women have to negotiate for the use of water, usually losing out, as men access it for irrigation, while women need it for maintaining livestock. Scheduled Caste women speak of appropriation of tanker water by upper-caste households and this finding accords with the increase in public water conflicts across India. We are increasingly becoming aware that when water shortages increase in climate-change situations, conflict over water will also increase (Mangi et al. 2019). Future research needs to both define and analyze the gendered impact of water conflict in climate-change situations. At the same time these conflicts are placing emphasis on new actors, such as middlemen who take advantage of these shortages, and this is an indicator of change linked to climate to which due attention must be paid by policy makers and researchers. The emergence of a private, informal water delivery sector undermines local sustainability and tends to exclude women-headed households. To face the challenge, a knowledge-based climate adaptation strategy can build women's capacity and contribute to their knowledge.

Work on migration and climate change has been increasing, with migration visualized as a distress response to climate change across the hotspots. Usually young men move out in search of livelihoods as they have to cope with crop failures, deteriorating pastures and increasing poverty. This results in an increase in women-headed households. Migration across South Asia is seen as a positive strategy (Adger et al. 2015; Afifi et al. 2016; Gemenne and Blocher 2017), with different patterns of short-term, seasonal and more longer-term migration dominated by economic drivers (Abbasi in this volume), yet it has gendered outcomes.

Household dynamics change "tangibly and intangibly" when men migrate (Hosegood et al. 2007; Nguyen and Locke 2014), as this leaves women behind as the sole carers and workers in the household. Women adapt by taking on male responsibilities in their absence. There is no indication whether workload benefits in the household are shared equally. Remittances from migrants do make a difference to the household, helping with poverty reduction and food security, as most remittances are spent on food

(Szabo et al. 2018). This is especially the case when migrants cross borders and send back foreign exchange, for example, in Bangladesh and India (Mahanadi Delta) (Vincent and Hazra in this volume).

Male migration needs more research, as vulnerability may not be confined to women, but also affects the men and the community. Migrating men face health issues as there is no health cover available when they work in the unorganized sector or face problems due to hazardous work (Udas et al. in this volume). Thus, climate change-linked migration requires new policy initiatives which are gendered and take into account the vulnerability of all groups.

While women have been adopting cash crops, hybrid seeds or mechanization (Qaisrani and Batool in this volume), seeking to build new skills for survival and strengthening livelihoods, they are often left out of technology change, or indeed finances to meet the new challenges. These inputs are usually identified as male needs and used as justifications for keeping women out of the fields of science, technology, engineering and maths (STEM) (Schmuck 2017). Women's work continues to be recognized as only within the household, even if they head households and engage with markets. Climate change and variability affect women-headed households more than others as their access to resources and knowledge is limited. Gendered vulnerabilities are thus determined by local factors such as exclusion from the market and technological interventions, together with the social structures of the household, the community and the state. Capacity building and improved finances become essential to their survival.

In analyzing the above issues, the authors have highlighted differentiated caste and class structures, and the implications for resource distribution and hence for policy and research (Udas et al. in this volume). It can be argued that adaptation policies and programmes that are responsive to community needs, seek to create an equal playing field. If this does not happen, then climate responses end up widening gender gaps. Across the hotspots, there can be no single response or solution to the impacts of climate change, as gendered distinctions are part of the varied factors in household–community interplay. While women may exercise individual agency, this does not necessarily remove the social vulnerability to which they are subject.

Expanding the fiscal and governance framework in South Asia

There is limited research on engendering climate change in South Asia (Bhatta et al. 2015; Rao et al. 2019). Recognizing vulnerability as a first step has been taken by all four countries in their climate change policies, but they have failed to go beyond this, as can be seen from the four countries' Nationally Determined Contributions to the UNFCC (WEDO 2016). The South Asia Association of Regional Countries (SAARC), which includes all the four countries in this volume, has a structure linked to climate change which started with the establishment of the SAARC Disaster Management Centre (SDMC) in 2006. Climate-related security risks above and beyond natural disasters were broadly emphasized in the 2007 Declaration of the 14th SAARC Summit. Though regional levels form the basis for the international and local, in South Asia, progress on cross-country policies has not been possible. Despite the several institutions that have been established by SAARC, they have not been able to produce concrete results and especially any gendered response or finance (Krampe and Swain 2018). The South Asian scholarship in the volume is vibrantly focused on the way gender is constructed as well as contested. All chapters, whether on India, Bangladesh, Pakistan or Nepal, address gender concerns in contexts of climate change and represent an important direction for the future. Women in South Asia carry out work within the home and in agriculture, are the main carers for the family, and in climate change situations also take on work that was traditionally done by men. By recognizing only the vulnerability of women and overlooking their agency, national policies are contributing to perpetuating a stereotypical image of defenceless women. Thus their capacity to mitigate the impacts of climate change is overlooked.

Though finance for climate change is part of the political debate at the international level, the targeted inclusion of women has yet to be initiated in South Asia. Policies on financial support for those who are affected and face survival issues requires budgets specifically designated for the gendering of climate change in the development economy. There is a need for increased income, grants, technology transfers and other forms of financial support which would require decisions at international and national level.

Some of the work carried out by women, such as crop rotation, crop substitution and shifting timings should be, but are not, part of any climate change funding (Williams 2016). Lack of finance may therefore have implications, as observed in the case of the drying up of borewells and women's assets, including gold, being spent on more borewells in the absence of cash. This disproportionately affects the Scheduled Castes and Tribes and poor households (Solomon and Rao in this volume).

Building capacities for gendered research in climate change: unpacking methodology and epistemology

Climate change linked to a new transformative agenda grounded in gendered concepts of equality requires research at each stage of systems change. However, systems vary in each country, hence the determinants of adaptive capacity also function differently in different contexts (Smit and Wandel 2006: 288; Moosa and Tuana 2014).

There is a need to build the capacities of researchers for designing, collecting and analyzing sex-disaggregated data related specifically to social processes and adaptive demands. Disaggregation, however, is not enough; deeper analyses of gender identities, power differentials, social relations and change are also required. For the future, studies must address the lack of a robust methodology to address gender inequalities from a multi-disciplinary perspective in a context of climate change. A distinguishing characteristic of this methodology has to be that it is an active engagement with women and men but must go beyond binaries and include identities of class, caste, community, and in the South Asian context, minorities. It should be able to explore missing as well as contextually situated research requirements of a 'socially engaged research' that interrogates existing knowledge while producing new knowledge based on ethical responsibility and making the reader aware of exclusion. With this in mind, through this volume we have used a diversity of methods ranging from participatory to qualitative and quantitative research to provide complex understandings, as authors have explored the interface between women and men across different social groups and positions, and climate change. Common to the research has been the use of critical methods to unpack a range of intersectionalities at work in women's everyday lives.

The epistemological position of women in climate change research needs to be based on women's lived experiences. This will enable future gendered research to view women and men through a critical socio-political lens. It will enable research and policy makers to go beyond the vulnerability paradigm and patriarchal structures to those that are inclusive of women's agency. An understanding of household dynamics shows women changing gender norms to control resources and remove cultural barriers. A challenging critique is thus making both women's and men's social worlds visible through the epistemic values of experiences.

There is a missing link between research and policy, and this empirical data could inform policy and development professionals to implement interventions that are gendered. Building climate resilience for both women and men cannot just be driven by mainstreaming gender in climate change policy; it requires women's needs and requirements to be integrated into all the other national and local policies and actions that shape and construct discriminatory gender roles in society and within households.

References

Adger, W., Arnell, N., Black, R., Dercon, S., Geddes, A., and Thomas, D. 2015. 'Focus on Environmental Risks and Migration: Causes and Consequences.' *Environmental Research Letters*, 10(6).

Afifi, T., Milan, A., Etzold, B., Schraven, B., Schulz Rademacher, C., Sakdapolrak, P., and Warner, K. 2016. 'Human Mobility in Response to Rainfall Variability: Opportunities for Migration a Successful Adaptation Strategy in Eight Case Studies.' *Migration and Development*, 5: 254–274.

Ahmed, S., and Fajber, E. 2009. 'Engendering Adaptation to Clime Variability in Gujarat. India.' *Gender and Development*, 17: 33–50. doi:10.1080/13552070802696896

Akter, S., Rutsaert, P., Luis, J., Htwe, N.M., Raharjo, S.B., and Pustika, A. 2017. 'Women's Empowerment and Gender Equity in Agriculture: A Different Perspective from Southeast Asia.' *Food Policy*, 69: 270–279. doi:10.1016/j.foodpol.2017.05.003.

Bee, B.A., Rice, J., and Trauger, A. 2015. 'A Feminist Approach to Climate Change Governance: Everyday and Intimate politics.' *Geography Compass*, 96): 339–350.

Bhatta, G.D., Aggarwal, P.K., Poudel, S., and Belgrave, D.A. 2015. 'Climate-induced Migration in South Asia: Migration Decisions and the Gender Dimensions of Adverse Climatic Events.' *Journal of Rural and Community Development*, 10: 1–23.

Chant, S. 1998. 'Households, Gender, and Rural–urban Migration: Reflections on Linkages and Considerations for Policy.' *Environment and Urbanization*, 10(1): 5–21.

Chowbey, P. 2017. 'Women's Narratives of Economic Abuse and Financial Strategies in Britain and South Asia.' *Psychology of Violence*, 7: 459–468.

Enloe, C. 2017. *The Big Push: Exposing and Challenging the Persistence of Patriarchy*. Oxford: Myriad Editions.

Fernandes, L., (ed.). 2018. *Routledge Handbook of Gender in South Asia*. London: Routledge.

Gemenne, F., and Blocher, J. 2017. 'How Can Migration Serve Adaptation to Climate Change? Challenges to Fleshing Out a Policy Deal.' *Geographical Journal*, 183: 336–347.

Gilligan, C., and Snider, N. 2018. *Why Does Patriarchy Persist?* Cambridge, UK: Polity Press.

Goh, A.H.X. 2012. '*A Literature Review of the Gender Differentiated Impacts of Climate Change on Women's and Men's Assets and Well Being in Developing Countries.*' CAPRi Working Paper No. 106. Washington, DC: International Food Policy Research Institute.

Goodrich, C.G., Udas, P.B., and Larrington-Spencer, H. 2019. 'Conceptualizing gendered vulnerability to climate change in the Hindu Kush Himalaya: Contextual conditions and drivers of change', *Environmental Development*, 31: 9–18. doi:10.1016/j.envdev.2018.11.003.

Hosegood, V., Preston-Whyte, E., Busza, J., Moitse, S., and Timaeus, I.M. 2007. 'Revealing the Full Extent of Households' Experiences of HIV and AIDS in Rural South Africa.' *Social Science & Medicine*, 65: 1249–1259.

Isran, S., and Isran, A.M. 2012. 'Patriarchy and Women in Pakistan: A Critical Analysis, Interdisciplinary.' *Journal of Contemporary Research in Business*, 4(6): 835–859.

Kabeer, N. 1997. 'Women, Wages and Intra-household Power Relations in Urban Bangladesh.' *Development and Change*, 28: 261–302.

Kandiyoti, D. 1988. 'Bargaining with Patriarchy.' *Gender & Society*, 2: 274–290.

Krampe, F., and Swain, A. 2018. 'Is SAARC prepared to combat climate change and its security risks?'. *The Third Pole Net*. https://www.thethirdpole.net/en/2018/09/06/is-saarc-prepared-to-combat-climate-change-and-its-security-risks/.

Lawson, V. 1998. 'Hierarchical Households and Gendered Migration: A Research Agenda.' *Progress in Human Geography*, 22(1): 32–53.

Mangi, F., Kay, C., Chaudhary, A. 2019. 'A Water Crisis Is Brewing Between South Asia's Arch-Rivals.' *Bloomberg*. https://www.bloomberg.com/news/features/2019-01-25/a-water-crisis-is-brewing-between-south-asia-s-arch-rivals.

Moosa, C.S., and Tuana, N. 2014. 'Mapping a Research Agenda Concerning Gender and Climate Change.' *Sociological Review*, 57: 124–140.

Ngigi, M., Mueller, U., Birner, R. 2017. 'Gender Differences in Climate Change Adaptation Strategies and Participation in Group Based Approaches: an Intra-household Analysis from Kenya.' *Ecological Economics*, 138: 99–108.

Nguyen, M.T., and Locke, C. 2014. 'Rural-urban migration in Vietnam and China: Gendered Householding, Production of Space and the State.' *Journal of Peasant Studies*, 41: 855–876.

Price, G., Alam, R., Hasan, S., Humayu, F., Kabir, M.H., Karki, C.S., Mittra, S., Saad, T., Saleem, M., Saran, S., Shakya, P.R., Snow, C., and Tuladhar, S. 2014. *Attitudes to Water in South Asia*. London: Chatham House Report.

Rao, N., Lawson, E.T., Raditloaneng, W.N., Solomon, D. and Angula, M.N. 2019. 'Gendered Vulnerabilities to Climate Change: Insights from the Semi-arid Regions of Africa and Asia.' *Climate and Development*, 11(1): 14–26. doi:10.1080/1756 5529.2017.1372266.

Rao, N., Singh, C., Solomon, D., Camfield, L., Sidiki, R., Angula, M., Poonacha, P., Sidibé, A., and Lawson, E. T. 2020. 'Managing risk, changing aspirations and household dynamics: Implications for wellbeing and adaptation in semi-arid Africa and India.' *World Development*, 125, 104667.

Schmuck, C. 2017. *Women in STEM Disciplines: The Y factor. 2016 Global Report on Gender in Science, Technology, Engineering and Mathematics*. Springer International Publishing Switzerland.

Seager, J., and Olson, A. 1986. *Women in the World: An International Atlas*. New York: Simon and Schuster.

Singh, C. 2019. 'Migration as a driver of changing household structures: implications for local livelihoods and adaptation.' *Migration and Development*, 8(3): 301–319. doi:10.1080/21632324.2019.1589073.

Smit, B., and Wandel, J. 2006. 'Adaptation, Adaptive Capacity and Vulnerability.' *Global Environmental Change*, 16(3): 282–292. doi:10.1016/j.gloenvcha.2006.03.008

Sultana, A. 2012. 'Patriarchy and Women's Subordination: A Theoretical Analysis.' *The Arts Faculty Journal*, 4(1): 1–18.

Szabo, S., Adger, W.N., and Matthews, Z. 2018. 'Home is Where the Money Goes: Migration Related Urban Rural Integrations in Delta Regions.' *Migration and Development*, 2324: 1–17.

WEDO. 2016. 'Gender and Climate Change: Analyses of Intended Nationally Determined Contributions.' https://wedo.org/wp-content/uploads/2016/11/WEDO_GenderINDCAnalysis-1.pdf.

White, S.C. 2017. 'Patriarchal Investments: Marriage, Dowry and the Political Economy of Development in Bangladesh.' *Journal of Contemporary Asia*, 47 (2): 247–272. doi:10.1080/00472336.2016.1239271.

Williams, M. 2016. *Gender and Climate Change Financing: Coming Out of the Margin*. London: Routledge.

Index

Page numbers in *italic* indicate figures. Page numbers in **bold** indicate tables.

Abbasi, Saqib Shakeel, 9, 85–102
action arena, 203–205, 213
adaptation and women-headed
    households, *187*, 190–191, **192–193**,
    193–194
Adaptation at Scale in Semi-Arid
    Regions (ASSAR) 2, 58n1
adaptive capacity 156–157, 178,
    232–233
adaptive strategies 8, 229–231
Adger, W.N. 44
Adivasi Janajati 122n11
agricultural labour 3, 33, 128, 130,
    132, 134, 136, 139, 143, 144, 175,
    208, 209
aicho paicho 120
Ama Chedingmo 118, 122n3
Anwar, Muhammad Zubair 85–102

Batool, Samavia 19–33, 34n3
Bazaz, Amir 58n1
Begum, Anwara 152–166
biophysical assets 204, **204**
Bio-Physical Vulnerability Index (BPVI)
    assessment 183, **184**, 186
borewells 7; climate variability 70;
    delayed failures 72; digging 70;
    drilling of 69; functional borewells
    73; micro-irrigation technologies 73;
    ownership 73; risk-ranking exercises
    71; Rural Development and
    Panchayat Raj Schemes 69; in South
    India 137–140; sowing-related issues
    71
Brockhaus, M. 220
Byg, A. 202

cash crops 65, 128, 130, 132, 133, 229,
    231
Chandni Singh 8, 58–79
Chauri 111, 122n6
Cheema 24
Chilime hydropower project 117
chogo system 115, 116, 120
Ciurean, R.L. 44
climate change funding 231–232
climate-change-linked migration
    8–9
climate hotspots 1, 10, 227
climate-induced floods, Bangladesh 1
climate migrants 155
climate refugees 155
climatic variability: gendered
    institutions *see* gendered institutions;
    in high-elevation site 212–213, **213**;
    in mid-elevation site **212**, 212–213;
    people's perception of 211; in plains
    211–212; prevalence of 209–211,
    210
Collaborative Adaptation Research
    Initiative in Africa and Asia
    (CARIAA) 2, 4, 58n1
collective action 5, 8, 9–10
communal water sources 146
community assets 204, **204**
Composite Vulnerability Index (CVI)
    185, 186, *186*, 200
contextual vulnerability 1

Das, Shouvik 172–196
de Campos, Ricardo Safra 152–166
Deltas, Vulnerability, Climate Change:
    Migration and Adaptation
    (DECCMA) project 2, 175, 196n6

Denton, F. 45
dhaki jhakri 120
Diaras, Gandak river basin in Bihar, India: adaptation policies and programmes 54; annual livelihood strategies 49; climatic parameters 38; data collection 38; early marriage 52; employment opportunities 50; female-headed households 48; flood zones in bihar 39, 40; Human Development Index 39; 2011 *India Human Development Report* 39; infrastructure development 52; land security 53; loan burden among households 48–49, 49; loss of property and mobile lifestyles 40; *machhan* 48; *Mahadalit* 47, 51; *Mushar* and *Chammar* castes 47; participatory assessment tools 38; patriarchal and regressive social practices 47; Pipara-Piprasi embankment 41, 42; *Pradhan Mantri Gramin Awas Yojana* 53; rainwater harvesting tools 53; role of institutions in times of flood 51–52; roof-based water-harvesting techniques 53; skill-based, non-farming castes 47; West Champaran *see* West Champaran, gendered vulnerabilities; *Yadav* 47
divisions of labour 132–135, *133*
Djoudi, H. 220
Douglas, M. 211
drip irrigation 58, 98, 101, 119, 134

Economic and Political Weekly (EPW) 6
Enloe, C. 227
environmental migration: climate-induced push factors 155; failure of adaptation 155–156; gendered effects of 156–157; human security 155; political implications 155; population movements and 154; role of 165
environmental stress 159
erosive/content migration 156

feminization of migration 154
flood irrigation 134

Focus Group Discussions (FGDs) 61, 62, 73, 74, 76, 88, 89, 92–94, 98, 107, 116, 130, 134, **206**, 207, 208
forced migration 160–161
Ford, J.D. 3
free electricity 128

Ganges-Brahmaputra-Meghna (GBM) delta *see* migration, Bangladesh
Geethalakshmi, V. 127
'gender-accepted' livelihoods 143
gender and climate change: Bhutan 5; caste and class 3; Collaborative Adaptation Research Initiative in Asia and Africa (CARIAA) 2, 4; diverse strategies and mechanisms 5; ethnic conflict, Northern Province 4; Family Background Report (FBR) 5; fishing 4; girls' education 4–5; Maldives 4; male migration 5; migration issues, Sri Lanka 4, 5; migration of women 8–9; natural resource-dependent sectors 2; negative effects on grain yields 3; policy interventions 4; sex-disaggregated data 4; Small Island Developing State (SIDS) 4
gender dynamics, Uttarakhand in Upper Ganga Basin (UGB): climatic variability 209–213; communities 202; data collection and analysis 205, **206**, 207, **208**; demographic profile 208–209; ethnic groups 205, **206**, 207; IAD framework *203*, 203–205, *204*; study area 207, *208*
gendered assets: gold 135–136; livestock 136–137
gendered institutions: accessing water for domestic use 213–215, **214**, *215*, 216, 217; agriculture activities 217, **218**, 219–220; dynamic, diverse and complex nature 220; intrinsic to 222; rules, norms and rights 220–221; social relations, position and mobility 221–222
Geographical Information System (GIS) 178
Ghale ethnic group 109, 112, 119, 120
Ghosh, Amit 172–196
Gilligan, C. 227, 228
Giri, Jasmine 172–196
Goodrich, Chanda G. 38–54

Govindan, Mini 201–223
Groundwater Regulation Bill 145

Habib, Nusrat 85–102
Hahn 19
Hans, Asha 1–11, 172–196, 226–233
Hazra, Sugata 172–196
Heyer, J. 132
high mountain communities, Upper
    Rasuwa in Gandak River Basin:
    average population density 107;
    basic facilities 106; bone-related
    problems 113; Chilime 110–112;
    class discrimination 119; climatic
    stressors 106; *Dalit* category
    109–110, 114, 119, 120; eco-agro
    tourism, home stay and handicrafts
    115–116; Gatlang 110, 112, 120;
    gendered vulnerabilities 107,
    111–112; Ghale ethnic group 109,
    112, 119, 120; Goljung 110;
    government intervention 120;
    Himalayan Adaptation, Water and
    Resilience (HI-AWARE) research
    project 107; household food security
    110; hydropower projects 111; *kami*
    group 110; *kharka* 110, 112, 113;
    migration, social capital, finance
    116–117; older people 119;
    psychology of 'untouchability,' 119;
    qualitative methods of data
    collection 107; reciprocal exchange
    of labour 114; satellite image 109,
    109; saving and credit activities 119;
    semi-structured checklist 107; snow
    cover and river discharge 107–108;
    social and economic position of
    women 114; spiritual and cultural
    practices 120; study area 107, *108*;
    Tamang ethnic groups 109, 119, 120;
    Task Allocation Study (TAS) 113,
    114; Tibeto-Burman Mongolian
    groups 109; tourism 111; trade and
    economic activities 117–118;
    transhumance herding, livestock and
    farming 112–114, **114**; water scarcity
    and competition 113; widows 119
Himalayan Adaptation, Water and
    Resilience (HI-AWARE) research
    project 2, 88, 107
Hindu Kush Himalayas (HKH)
    201
Hindu Succession Act of 1956 143

household cooperation 139–142, *140*
Hugo, G.J. 153
Human Development Index 39
hydrogeological features 138, 143
Hyndman, J. 4, 88

IAD framework *see* Institutional
    Analysis and Development (IAD)
    framework
indebtedness 137–139
in-depth interviews 130
2011 *India Human Development
    Report* 39
in-situ livelihood adaptations
    165
Institutional Analysis and Development
    (IAD) framework: action arena
    203–205; biophysical and
    community assets 204, **204**;
    collective choice rules 205;
    constitutional choice rules 205;
    operational rules 205
Intergovernmental Panel on Climate
    Change (IPCC) 178, 191
inter-household relations: geographical
    locations 227–229; in research
    226–227
intersectionality 9–10
intersections of geography and social
    identity 6–8
intra-household relations 144, **144**;
    geographical locations 227–229; in
    research 226–227
Islamic Shariah law 100

Janai Purnima 118

Kashif, M. 34n2
Khandekar, Neha 201–223
Khan, Qaiser 85–102
KNMI Climate Explorer 24

labour scarcity 129
land and well ownership 130, **131**, 132
Lázár, Attila N. 152–166
local women vulnerabilities and
    resilience, in Indus basin 98, **99**;
    adaptation strategies 89, 100; annual
    precipitation ranges 85; biologic
    vulnerability 101; climate change
    86–87, 89; climate-smart
    interventions 102; climate variability

and water availability 86, 101; cultural barriers 87; disaster risk reduction (DRR) 102; in downstream basin 97, 97–98, 100, 101; focus group discussions (FGDs) 88, 89; gender and climate change 86–87; high rainfall zone (Tehsil Murree) 94–95, 95; Himalayan Adaptation, Water and Resilience (HI-AWARE) Intervention Sites 88; Hunza basin 88; Islamic Shariah law 100; low rainfall zone (Tehsil Murree) 88, 95–96, 96; medium rainfall zone (Chakri) 95, 96; in mid-stream basin 93–94, 98, 100, 101; migration 87; North-West Frontier province (NWFP) 85; rural customs and urban norms 100; Sindh province 85; snowfall 85; Soan basin 88; social barriers 100; socio-economic situation 89, 90–91; in upstream basin 89, 92, 92–93, 98, 101

Mahatma Gandhi National Rural Employment Guarantee Act (MGNREGA) 134, 139, 176
male migration 5, 6, 9, 115, 116, 178, 179, 182, 187, 231
mani rimdu 120
Megh Pyyne Abhiyan 53, 54n2
men-headed households, Mahanadi Delta (MD): coastal hazard and migration 176, 178–179, 179; distribution of households 187, 187; employment opportunities 189; household survey 189, 190; patterns of 187, 188; socio-economic characteristics 180, 181–182, 182; study area 175, 175–176, 177; vulnerability 182–183, 184–185, 186, 186
migrant households 9, 159, 161–164, 162–164, 180
migration, Bangladesh 152; commitment 153; displacement 160–161; environmental change and 154–157; future migration flows 164–166; gendered patterns 153–154, 157–160, 158, 160; household survey 157; internal migration 159–160; international migration 158–160; migrant-sending areas 161–164, 161–164; structural-functionalist approach 153

Mueller, V. 24

National Action Plan on Climate Change (NAPCC) 174, 196n5
National Climate Change Policy 33
National Disaster Management Authority 209
National Family Health Survey (NFHS) 174, 195n3
neerganti 68, 77
New Economics of Labour Migration theory 153
non-migrant households 162–164, 162–164, 180
North, D. 202
north-east monsoon 132, 137
North-West Frontier province (NWFP) 85

Odisha 2010–15 Climate Change Action Plan (OCCAP) 191
off-farm employment 134
on-farm labour 134
organizational space 10

Palanisamy 138
Panthi 19
parma 120
participatory rural appraisal (PRA) tools 205
Patel, Amrita 1–11, 172–196, 226–233
Pathways to Resilience in Semi-arid Economies (PRISE) 2, 34n1
patriarchal bargain 11
Pradhan Mantri Gramin Awas Yojana 53
Prakash, Anjal 1–11, 38–54, 226–233
"privilege, power and geography," 227
Public Distribution System (PDS) 135, 147n4

Qaisrani, Ayesha 19–33, 34n2

raashan 51, 52
Rao, Nitya 1–11, 58n1, 127–147, 226–233
Rathod, Roshan 201–223

remittances 9, *161*, 161–162
Rijhwani, Vani 201–223
rimthim 120
Rural Development and Panchayat Raj
　schemes 69
Rural Poverty Alleviation Project 115
rural women vulnerabilities, in Dera
　Ghazi Khan, Faisalabad: class
　matters 26–30; cotton production
　28; cotton value chains 20;
　desertification and land degradation
　22; food insecurity situation 22;
　household survey 20; inequity traps
　19; inter-generational dynamics and
　decision making 30–32;
　Intergovernmental Panel on Climate
　Change (IPCC) Fifth Assessment
　Report (AR5) 22; labour force
　participation rate 25; lower-and
　middle-income groups 29;
　multidimensional poverty rate 25;
　natural resource-based livelihoods
　23; PRISE projects 20; risk
　developing skin allergies 28; sample
　selection by district 20, **21**;
　socioeconomic indicators 23, **23**, 25;
　socio-economic status 25;
　temperature trends 24, *24*, 25; Union
　Councils (UCs) 20; urban–rural ratio
　24; water availability 22;
　waterlogging and salinity issues 23;
　wheat production 28; Women's
　Economic and Social Empowerment
　(WESE) Index 29

Saeed 34n3
salgar 120
Salick, J. 202
Self-Help Groups (SHGs) 221
sensitivity 178
Sharma, Divya 201–223
simple arithmetic mean 200
small ruminants 130
Snider, N. 227, 228
'socially engaged research,' 233
Socio-Economic Caste Census (SECC)
　174
Socio-Economic Vulnerability Index
　(SEVI) assessment 183, **185**, 186
Solomon, Divya Susan 7, 58n1,
　127–147
South Asia Association of Regional
　Countries (SAARC) 231–232
south-west monsoon 62, 132
Sri Sathyan, B.N. 69

stratified proportional sampling
　129–130
surface water 128–129

Tamang, Deepak Dorje 106–122
Tamang ethnic groups 109, 119, 120
Tamil Nadu Minor Irrigation Scheme
　138
Tamil Nadu Panchayats Rules 145–146
transnational mobility 154
trapped population 187

Udas, Pranita Bhushan 38–54, 106–122

Vincent, Katharine 9, 152–166

water access and implications, Kolar:
　agrarian transitions 78; agriculture
　61; bicycles, informalization of
　drinking water access 73, 73–77, *74*,
　75; biophysical shifts 75; block
　chosen **84**; block-wise rainfall
　patterns 63, *64*; borewells 69–74;
　cash crop 65; climate-smart practices
　58; climate variability 61; cropping
　patterns and practices 65, *66*; details
　of villages **84**; drought, water scarcity
　and climate variability 62–63, *63*;
　ecological and climatic changes 59;
　environmental change 78;
　groundwater extraction 61; growing
　informalization of drinking water
　access 60; household survey 62;
　individualization of irrigation water
　60; informal interventions 59;
　informalization and fragmentation
　78; integrated water management 58;
　kere system 68–69, 79n4;
　landholding and caste 64; land use
　and cropping patterns 67, *67–68*;
　main employment categories 63, *65*;
　mixed-methods approach 61–62;
　natural resource degradation 78;
　neerganti 77; privatization of
　drinking water 58; reactive
　approaches 78; settlement scale 78;
　social-ecological system 59, 60;
　social-ecological system level 79;
　water availability 58; water
　collection for domestic purposes 79;
　watershed development 58
water-intensive cash crops 128, 130

water scarcity 127–129
weighted arithmetic mean 200
wells and well-being, South India:
  borewells 137–139; context and
  methodology 128–132, *129*, **131**;
  cropping patterns 132–135, *133*;
  divisions of labour 132–135, *133*;
  female-headed households 143–144;
  gendered assets 135–137;
  groundwater and gendered wellbeing **144**, 144–146; household
  cooperation, conflict and decision
  making 139–142, *140*; indebtedness
  137–139; socio-economic impacts
  128; water scarcity 127–128
West Champaran, gendered
  vulnerabilities 39, 40; anaemia 43;
  caste system 43; climatic modelling
  data 43; daily-wage work 47;
  differential gender impact 45, *45*;
  gender-based dynamics 45; gender-
  based vulnerability 44, 46–47;
  gendered life 44–45; gender
  inequalities 45; Intergovernmental
  Panel on Climate Change (IPCC) 43;
  land ownership 41, *42*; literacy rate
  41; 2015–16 National Health Survey
  41; Representative Concentration
  Pathways (RCP) scenario 43; social
  vulnerability 44; socio-economic
  inequality 41; top down *vs.* bottom
  up approach 44, *44*; *zamindari*
  system 41

widow-headed households, Mahanadi
  Delta (MD): socio-economic
  characteristics 180, **181**–182, 182;
  study area *175*, 175–176, **177**;
  vulnerability 182–183, **184–185**,
  186, *186*
Wildavsky, A. 211
Wolfe, A.K. 211
women-headed households, Mahanadi
  Delta (MD): adaptation *187*,
  190–191, **192–193**, 193–194; Census
  of India 174; coastal hazards risk
  176, 178–179, *179*; household
  survey **177**, 179–180; investigation
  of 173; migration 176, 178–179,
  179, 187, *187*, **188**, 189, *190*;
  population density 175; SECC 174;
  socio-economic characteristics 180,
  181–182, 182; study area *175*,
  175–176; vulnerability 182–183,
  184–185, 186, *186*; work
  participation rate
  175–176
Women's Economic and Social
  Empowerment (WESE)
  Index 29

zamindari system 41
Zwarteveen, M. 139

Printed in the United States
by Bookmasters

Printed in the United States
By Bookmasters